Foam Fractionation

Principles and Process Design

Foam Fractionation

Principles and Process Design

Paul Stevenson and Xueliang Li

CRC Press
Taylor & Francis Group
Boca Raton London New York

CRC Press is an imprint of the
Taylor & Francis Group, an **informa** business

CRC Press
Taylor & Francis Group
6000 Broken Sound Parkway NW, Suite 300
Boca Raton, FL 33487-2742

First issued in paperback 2017

ISBN-13: 978-1-4665-5851-9 (hbk)
ISBN-13: 978-1-138-07428-6 (pbk)

This book contains information obtained from authentic and highly regarded sources. Reasonable efforts have been made to publish reliable data and information, but the author and publisher cannot assume responsibility for the validity of all materials or the consequences of their use. The authors and publishers have attempted to trace the copyright holders of all material reproduced in this publication and apologize to copyright holders if permission to publish in this form has not been obtained. If any copyright material has not been acknowledged please write and let us know so we may rectify in any future reprint.

Visit the Taylor & Francis Web site at
http://www.taylorandfrancis.com

and the CRC Press Web site at
http://www.crcpress.com

Contents

Preface

As far as we are aware, this book represents the first monograph on the process of foam fractionation. This might not be particularly surprising to some. Despite extensive work on foam fractionation in research laboratories, it has yet to gain traction in industry despite its demonstrable process potential. We hope that the book will help rectify this situation by introducing the reader to what we regard as the underlying physical basis for the process, giving design advice and discussing a case study of one of the few current commercial applications of the process. As such, the book is primarily aimed at practicing process engineers who may be in need of support and reassurance when contemplating which simple and cheap separation process for surface-active molecules from liquid solution should be installed in their plants. There are disproportionately more academics engaged in research into the process than there are practitioners practicing it; it is hoped that such researchers might find fertile areas of research within which they can ultimately promote widespread industrial adoption. Foam fractionation lies at the nexus of the disciplines of "multiphase flow" and "surface science," with a little bit of "equilibrium stage processes" included, and the material presented herein will be of interest to some senior undergraduate and graduate students in process/chemical engineering.

It is appropriate to say something about how we came to work with foam fractionation, and with each other: Paul Stevenson (PS) was working on foam flowing up vertical columns at the University of Newcastle (Australia), ostensibly in the context of froth flotation. He applied for a job with Professor Richard Darton at the University of Oxford, who was working extensively on the process at the time. PS's application was unsuccessful, and instead his friend Dr. Peter Martin, who was at the University of Manchester at the time (and subsequently returned there), got the Oxford job. It was Martin who substantively introduced PS to foam fractionation, and discussions continued when Martin traveled to Australia to attend PS's wedding.

Around the same time, Xueliang Li (XL) was completing his Master's studies in biochemical engineering at Hebei University of Technology, Tianjin, China. His research project was on foam fractionation, and was specifically supporting the foam fractionation process at the Tianjin Kangyi Biotechnology Co., Ltd., who were producing nisin and employed what is, to the best of our knowledge, the largest foam fractionation installation in the world. XL worked on a part-time basis at the factory and was partly responsible for its design, commissioning, and early operation.

We discovered one another online, and communicated extensively online for several months until PS won substantial support from the Australian

Research Council that enabled him to invite XL to Australia to commence doctoral studies in "Transport Phenomena in Foam Fractionation." Much of this book is based on the results of the subsequent collaboration, and on XL's doctoral thesis.

Paul Stevenson and Xueliang Li

Acknowledgments

The authors would like to thank the staff at CRC Press, especially Allison Shatkin and Laurie Schlags, for initially considering our proposal and subsequently providing excellent support throughout the writing process. We would like to thank Professors Bob Tanner and Denis Weaire, and Dr. Peter Martin for conversations about foam and foam fractionation over the years. We are grateful to Noel Lambert for discussions about the parallels that exist between froth floation and foam fractionation. Thanks go to Sven Schröter, who took the photograph that appears on the front cover, and to Fiona Childs of the University of Hull for administrative support. PS would like to thank the China University of Petroleum (East China), and especially its Vice-President, Professor Zhaomin Li, for provision of magnificent facilities in which to write much of this book while working there in the role of Visiting Professor. XL would like to thank Tianjin Kangyi Biotechnology Co., Ltd. for stimulating his interest in the foam fractionation technology, and for providing the industrial experience and perspective that have contributed to his doctoral research, and the case study detailed in Chapter 10 herein. However, the biggest thanks go to the readers of this book for sharing an interest in the process that both of us have been in love with for several years, although it is likely that our wives, Tracey and Xinting, were significantly less enamored with it. Although neither of us is currently engaged in foam fractionation practice or research, it is quite certain that we will eventually return.

About the Authors

Paul Stevenson is a lecturer in chemical engineering at the University of Hull, guest professor at the China University of Petroleum (East China) and director of Lower Belford Resources. He has previously held academic positions at the Universities of Auckland and Newcastle (Australia), and undertook post-doctoral training at the University of Cambridge, the latter at which he had previously taken four degrees in chemical engineering. Stevenson is Visiting Professor at the China University of Petroleum (East China) and Adjunct Associate Professor in the Reservoir Engineering Group of the University of Western Australia. His research work lies at the nexus of multiphase flow and surface science, and currently involves investigations into enhanced oil recovery, gas–liquid mass transfer, and mineral flotation. Before embarking on an academic career, Stevenson was a trader of Japanese derivative securities and an independent racecourse bookmaker.

Xueliang (Bruce) Li received a Master's degree from the Department of Bioengineering, Hebei University of Technology, China, where he researched the foam fractionation process in collaboration with Tianjin Kangyi Biotechnology Co., Ltd., and subsequently worked with this company in 2007. He participated in the design of the only industrial protein foam fractionation units in the world, with a throughput of 30 tons per day. During this period, he regularly discussed fosam fractionation with Dr. Paul Stevenson and Professor Geoff Evans and, in May 2008, commenced doctoral studies at the University of Newcastle, Australia. Into the second year of his doctoral study, Li moved to the University of Auckland, New Zealand, and worked as a research assistant in the Department of Chemical and Materials Engineering, where he delivered tutorials and laboratory demonstrations in a wide range of second-year undergraduate courses. He also co-supervised several fourth-year undergraduate research projects. Upon graduation in 2012, Li joined LanzaTech NZ Ltd. as a development engineer and he currently leads the bioreactor scaling-up effort.

Li's research has focused on mid- to downstream bioengineering processes, including fermentation and product recovery. He has publications in international journals that cover the fields of applied microbiology, interface science, multiphase flow phenomena, and computer simulation of chemical processes. He is also an amateur programmer, and the apps he has developed are used by millions of users around the world.

1

Introduction

1.1 Self-Cleaning Fish Tanks

Look into a freshwater aquarium and you will almost always see air bubbles coursing through the water from the bottom of the tank to the top. To the non-specialist these might appear to perform merely an oxygenation role, which, of course, is essential for piscine life. However, there is often, but not always, another function that helps to clear the water of detritus originating from fish and aquatic plants. Such units are known as "protein skimmers."

The function of protein skimmers is fully described by Escobal (2000) but for the purposes of this introduction, a brief qualitative description of a typical unit will suffice. Air is compressed and "sparged" (i.e., introduced as bubbles) to a "riser" (i.e., a vertical tube), into which there can be ingress of tank water. Proteins, amino acids, fatty acids, and fats originating from the resident fish, as well as from certain plants, exist within the water and must be either periodically or continuously removed. Because some of these molecules possess hydrophobic groups, they tend to adsorb (i.e., attach to the surface of) the bubbles as they travel up the riser. These bubbles might aggregate to form a foam toward the top of the column that discharges over a weir, taking with it lots of gas–liquid surface, which is where the unwanted detritus has collected, and only a little bit of water. Thus, the tank water is continuously cleaned while the continuous loss of liquid itself is low.

It is difficult to know how many protein skimmers are currently in operation globally. According to the American Pet Products Association National Pet Owners Survey for 2011–2012, 11.9 million households with freshwater aquaria in the United States kept a total of 151.1 million freshwater fish. Although only a minority of the more sophisticated piscine hobbyists will run protein skimmers, it is possible that the number of units currently running stretches into the millions in the United States alone.

Protein skimmers are a subclass of the multiphase separation unit operation known as "foam fractionation" that works by having molecular species adsorb onto bubble surfaces. There are many more applications than simply cleaning fish tanks, and it is the process of foam fractionation, as well as the very closely related processes of ion flotation and bubble fractionation, that is the subject of this book. However, it must be stressed that the number of industrial processes that currently adopt the technique of foam fractionation is limited indeed, for reasons that will be expanded upon later in this book. Foam fractionation is itself a subclass of multiphase separation processes. The processes of column distillation, froth flotation, and ion exchange are multiphase separation processes too; and because they are widely adopted in industry, they are well developed and well known. Certainly they have had much greater penetration into the psyche of the general chemical engineer than has foam fractionation, but this is possibly unfair because, due to the existence of protein skimmers in aquaria, it is possible that, in fact, foam fractionation is the most common multiphase separation process in the world.

If the description of self-cleaning fish tanks using protein skimmers as an accessible example of foam fractionation is seen as too eclectic, one might like to consider another very common, although inadvertent, multiphase separation process that concentrates proteins in the foam layer (or "head") present on the top of beer. Bubbles rise through the beer, and, as they pass up through the liquid, proteins tend to adsorb onto their surface whence the bubbles form a foam layer at the top of the beverage. The protein content of the foam layer is typically found to be around 73% (w/w) with less than 10% (w/w) water, while the beer below has a protein concentration of merely 0.3% (w/w), according to Couper et al. (2010). This means that the bubbles have enriched the protein concentration in the liquid contained in the foam by an enormous factor of around 240. Therefore, when drinking a pint of the Irish stout Guinness that has a very tenacious head (see Section 4.2 on disproportionation in foam), it is nutritionally expedient to ensure that the froth remaining at the bottom of the glass is consumed.

1.2 The Principles of Foam Fractionation

The process of foam fractionation occurs precisely as described in Section 1.1 for protein skimming in that the following two transport phenomena occur:

1. Amphipathic molecules (i.e., those possessing both hydrophilic and hydrophobic groups) adsorb to the surface of gas bubbles as they pass through liquid, and
2. The bubbles are removed from the system as a foam.

The reason that foam is a good medium for adsorptive separation is that it has a very high specific surface area (i.e., the surface area per unit volume of foam) to which the target species adsorb, and the amount of interstitial liquid is low, particularly if the foam is very dry. It is the surface rather than the interstitial liquid that drives the separation process because the target species tend to journey from the interstitial liquid to the surface, and therefore the surface must be harvested, but the interstitial liquid is rejected as far as is possible. Indeed, the authors' present mantra in designing foam fractionators is that

> "The gas–liquid surface is the foam fractionator's friend, but interstitial liquid is his enemy."

Foam fractionators can be operated in (1) batch, (2) semi-batch, or (3) continuous mode, as will be described in Chapter 6. In the simplest, but generally commercially impractical, batch operation, foam is initially generated up to a desired height by passing gas into the liquid pool through a sparger. Then the gas supply is stopped, the foam is allowed to drain, and the liquid fraction of the foam decreases as drainage proceeds. The average bubble size will also change due to inter-bubble gas diffusion and rupture of foam lamellae, although the rate may be very low. After a particular time of drainage, the gas is turned on again and the newly generated foam drives the dry foam out of the column, and a new cycle of foam generation and drainage takes place. The foam is collapsed mechanically or chemically using a defoamer, and the resulting liquid, enriched in surface-active species, is called the foamate.

The concentration of the surface-active material in the foamate is a function of the specific surface area of the foam (in units of $m^2.m^{-3}$) and the surface excess (i.e., the surface concentration, in units of $mol.m^{-2}$). By assuming that the surfactant concentration in the interstitial liquid of the foam (see Figure 1.1) is the same as the surfactant concentration in the liquid pool, the enrichment ratio can be calculated by

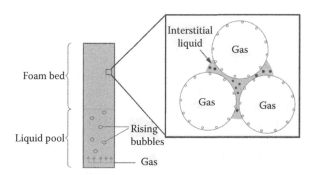

FIGURE 1.1
Schematic diagram of a simple foam fractionator. Inset: schematic diagram of surfactant molecules adsorbed at the interface (○) and dissolved in the interstitial liquid (●).

$$E = \frac{C_f}{C_b} = 1 + \frac{\Gamma S}{C_b} \tag{1.1}$$

where C_f (mol.m^{-3}) is foamate concentration and C_b (mol.m^{-3}) is the bulk concentration; Γ (mol.m^{-2}) is the surface excess. S (m^{-1}) is the specific surface area of the foam given, as demonstrated by Stevenson (2012), by

$$S = \frac{3(1 - \varepsilon)}{r_{32}} \tag{1.2}$$

where ε(-) is the volumetric liquid fraction of the foam and r_{32} (m) is the Sauter mean bubble radius defined statistically by

$$r_{32} = \frac{\displaystyle\int_0^\infty r_b^3 f(r_b) \, \mathrm{d}r_b}{\displaystyle\int_0^\infty r_b^2 f(r_b) \, \mathrm{d}r_b} \tag{1.3}$$

where $f(r_b)$ is the probability density function of the given bubble size distribution. For instance, the equilibrium surface excess (a term that will be further defined in Section 2.2) for a 5×10^{-3} mol.m^{-3} solution of the nonionic surfactant $C_{10}E_8$ is about 1.7×10^{-6} mol.m^{-2} (Yang and Gu, 2004). If a foam consisting of bubbles of 1 mm Sauter mean radius and a liquid fraction of 1% is used to enrich this solution, the resultant enrichment ratio will be about 100. In such a foam, the specific surfaces area is 2,970 m^2.m^{-3}, which compares very favorably with the figures for other gas–liquid contacting devices given by Charpentier (1981); for a bubble column, for instance, a specific surface area of 50 to 600 m^2.m^{-3} was given.

In the above discussion, batch mode operation is used as an example. In a batch mode operation, the foam is essentially static so that the enrichment can be simply calculated from Equation 1.1. However, in a rising column of foam, such as those employed in semi-batch (continuous with respect to the gas phase, batch with respect to the liquid) or continuous (continuous with respect to both phases) operations, the liquid fraction of the foam in the column is not directly related to the final enrichment of the process due to the slip velocity between the liquid and gas phases. A detailed discussion on the implication of the existence of a slip velocity is presented in Chapter 3. Despite its great simplicity, batch operation has little practical utility because of its low time efficiency (the system is idle during the drainage phase) in comparison with semi-batch and continuous operations. Batch operation is not suitable for studying the hydrodynamics of foam either, because of its temporal and spatial variance. A true steady state can only be achieved in a continuous operation.

1.3 Relationship of Foam Fractionation to Froth Flotation

Because foam fractionation is so very similar to the more established process of froth flotation, it is pertinent to discuss the latter in this introduction, because knowledge of flotation will be applied to foam fractionation throughout this book. Froth flotation is a process that is used in coal preparation and mineral beneficiation to separate the valuable coal or minerals from the unwanted "ash" or "gangue" material. The mechanism of separation is based on differences in the hydrophobicity of the particle species. The simplest explanation of flotation is as follows: It is not energetically favourable for hydrophobic particles to reside wholly within a liquid, and therefore, given the opportunity, they will attach bubbles and air and will 'hitch a ride' with the bubbles up the column whereas the more hydrophilic particles will not attach and will instead fall in a column.

This is not a book about froth flotation. The interested reader is invited to consult Nguyen and Schulze (2003) for an excellent text on the colloidal basis of flotation, and Finch and Dobby (1990) for an introductory text on column flotation. However, froth flotation is such a close cousin of the process of foam fractionation that a brief consideration of its operation is appropriate herein, especially because there has been an enormous research and development effort in the field of froth flotation (2005 saw celebrations for the centenary of industrial application of flotation in Australia; Fuerstenau et al., 2007), whereas foam fractionation is very much a nascent technology. An understanding of the principles and process innovation of froth flotation can provide the engineer who is trying to develop a foam fractionation process with an excellent head start.

There is typically a deep foam layer in a flotation column, whereas a mechanical flotation cell often supports very little foam indeed. Thus, a consideration of column flotation is more pertinent if we seek analogies with foam fractionation. Figure 1.2 shows a schematic representation of a flotation column, after Finch and Dobby (1990). A slurry of particles is suspended in water to the column in the feed stream. In the case of the flotation of coal, these particles consist of some that are predominately coal and some that are predominately "ash" (i.e., material that will not burn). In the case of the flotation of coal, there are two reagents that are added to the feed:

1. A "frother," which is a surface-active molecule, typically methyl isobutyl carbinol (MIBC) or a proprietary frother in coal flotation, that acts to adsorb to the gas–liquid interface to reduce the surface energy and therefore stabilize the froth, and

2. A "collector," which, in coal flotation, is typically diesel fuel in small quantities that preferentially adsorbs to the coal surface to enhance its hydrophobicity.

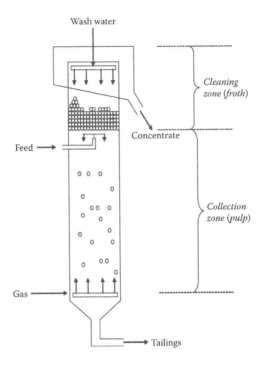

FIGURE 1.2
A schematic representation of a column flotation unit. (Adapted from Finch, J.A., and Dobby, G.S. 1990. *Column Flotation*, Pergamon Press, ISBN 0080401864.)

In minerals flotation, there is the potential to add a further two reagents:

3. An "activator," which helps the collector selectively adsorb to a particular particle species, and
4. A "depressant," which helps prevent the collector from adsorbing to a particular particle species.

When air is bubbled through the slurry in the "collection zone," the coal particles have the opportunity to attach to the bubbles, whence they are transported up the column and into the foam layer above, known as the cleaning zone. The hydrophilic ash particles do not attach and most fall through the slurry to be rejected via the bottom (or "tailings") stream. Some ash particles do get entrained in the foam layer, but "washwater" added to the surface of the froth ensures that most entrained ash is rejected from the foam layer.

Although column flotation is specifically discussed herein, all flotation devices work by separating particles based on differences in their hydrophobicity. Hydrophobic solid surfaces dislike contact with water, and so they will endeavor to attach to a bubble so as to form more solid-gas interface, which is energetically favorable. An example of a hydrophobic surface is the performance fabric Gore-Tex™. Because it is hydrophobic, the fabric is not well-wetted by the water and a high "three-phase contact angle" is exhibited.

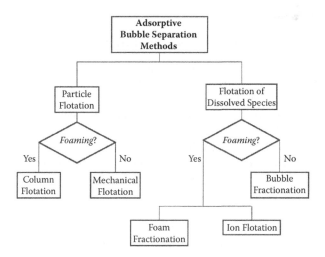

FIGURE 1.3
A hierarchy of adsorptive bubble separation methods. The processes considered herein are the ones on the right-hand side.

On the other hand, it is energetically favorable for a hydrophilic surface to water-wet and therefore it exhibits a relatively low three-phase contact angle.

The three-phase contact angle (i.e., how hydrophobic or hydrophilic a surface is) is determined by two factors: (1) the physico-chemical nature of the solid material, which is something that we need not dwell upon; and (2) the surface roughness of the solid surface, which we will dwell upon because it is a dependency so often overlooked, and, even though there is no direct relevance to foam fractionation, it is a dependency that cannot be over-stressed to the flotation community. The three-phase contact angle of a water droplet upon a perfectly smooth surface is determined by physico-chemical properties only, and this interaction might be either relatively hydrophobic or hydrophilic. However, any roughness elements apparent on the surface have the effect of making hydrophobic surfaces still more hydrophobic but hydrophilic surfaces still more hydrophilic through a phenomenon known as the Wenzel Effect (Wenzel, 1949). We note that the roughness caused by the weave of Gore-Tex™ fabric is likely to enhance the hydrophobicity of the surface.

Thus, whether or not a particle wants to attach to a bubble is determined by its hydrophobicity. Once a hydrophobic particle attaches (and if it remains attached) in the collection zone, it is transported to the cleaning zone where it travels through the froth still attached to the bubble surface, whence it travels over a weir in the "launder" and is recovered in the "concentrate" stream.

It will be seen that the general principles of froth flotation are almost identical to those that underpin foam fractionation, save for the fact that in the former the attachment of hydrophobic particles to bubble surfaces is the fundamental physical mechanism, whereas in the latter it is the adsorption of amphipathic molecules to gas–liquid interfaces.

1.4 A Brief History of Process Development

It is difficult to trace the exact genesis of foam fractionation. Charm (1972) suggests that one of the earliest reports of the process was due to Ostwald and Siehr, who separated proteins from a waste stream in the manufacturing of starch for use in animal feed, which was a process subsequently studied by Prokop and Tanner (1993). Charm cites a slew of literature published subsequently into the early 1950s of the use of foam fractionation for the recovery of enzymes and other proteins, although the uptake of the process by industry is not clear. However, undoubtedly the father of modern investigations into the process of foam fractionation was Robert Lemlich of the University of Cincinnati. Lemlich recognized that true process design, and innovation therein, could only come from an understanding of the physics that underpin the process. He therefore spent much of his career not only researching foam fractionation, as well as methods of process intensification, but also engaged in investigations into the fundamentals of gas–liquid foam behavior. It is precisely in the spirit of Robert Lemlich that the preparation of this book has been attempted.

Lemlich commenced his work on foam fractionation by publishing an investigation of a separation column with external reflux (Lemlich and Lavi, 1961), which will be discussed in Section 6.4, and continued process development work for the next three decades. However, supporting such development was some excellent work on the fundamentals of gas–liquid foam; the present authors occasionally joke with each other that if there is a novel development of experimental equipment to investigate foam, or a clever new (and seemingly correct) theory reported in the literature, it is worth checking to see if Robert Lemlich did it 50 years ago. A good example of Lemlich's contribution to the fundamental physics of gas–liquid foam is work that he published in 1965 to relate the process of foam fractionation to the rate of liquid drainage in foam. The difficulty was that there was no mechanistic description of foam drainage. Leonard and Lemlich (1965a) first calculated the velocity profile of a liquid flowing in a channel of a cross-sectional shape of a convex triscupid, which is approximately the shape of "Plateau border" channels in foam, before using this result to obtain a description of liquid drainage in foam in the presence of capillary effects (Leonard and Lemlich, 1965b). This work was subsequently supported by an extensive experimental investigation to relate the theoretical results to foam fractionation practice (Leonard and Lemlich, 1965c). Although it may be stretching the bounds of credibility to describe the year 1965 as the *Annus mirabilis* of foam fractionation, it certainly did mark an outstanding contribution to both the theory and practice of the process. In trying to physically describe the minority process of foam fractionation, Lemlich had mathematically described the much more general physical phenomenon of foam drainage.

It is therefore unfortunate that Leonard and Lemlich's drainage theory subsequently was largely overlooked. In the USSR, Gol'dfarb (1988) published an identical foam drainage theory seemingly unaware of Lemlich's work, as did Weaire (Verbist et al., 1996). Both later works had to assume that the gas–liquid interfaces were infinitely rigid, whereas Leonard and Lemlich were significantly more sophisticated by allowing for finite rigidity (via the surface shear viscosity). The Verbist drainage model has become established in the canon of gas–liquid foam as "The Foam Drainage Equation." Neither of the present authors ever had the privilege of meeting Professor Lemlich but we do not think that he would have particularly minded this injustice.

Lemlich (1972) edited a volume titled *Adsorptive Bubble Separation Processes*, in which various authors wrote chapters on foam fractionation, or its allied processes of flotation, ion flotation, and bubble fractionation, with one chapter by de Vries (1972) being devoted to the fundamental behavior of gas–liquid foam, and especially important considerations of bubble size distribution. The relationship of foam fractionation to its allied processes is shown schematically in Figure 1.3. The de Vries chapter, along with excellent contributions from Lemlich himself, aside, the volume could be said to be disappointing because it exhibits a great deal of repetition, which is arguably inevitable in an edited volume with multiple authors on a specialized topic. The present authors would like to think that this monograph represents a significant advance on Lemlich's volume, not only because there have been 40 years of subsequent research and development on foam fractionation, but also because an attempt has been made to present it in a coherent and consistent fashion. Lemlich continued to develop his ideas about gas–liquid foam in general, and foam fractionation in particular, until early 1989 when he published the results of a study on foam under simulated conditions of microgravity (Cengel and Lemlich, 1989), some 15 years before the foam experiments in parabolic flight of Saint-Jalmes et al. (2006). It is typical of how Lemlich's work has been overlooked that no author writing about foams in microgravity has ever cited the pioneering work in the field. As far as the present authors know, at the end of Lemlich's productive research career, there was still not a major industrial foam fractionation operation in operation anywhere in the world.

After Lemlich, research and development into the fundamental physics of foam fractionation arguably experienced a hiatus. There have been a great many studies showing that, for instance, one protein or another can be stripped from solution using foam fractionation, but no progress in suggesting general design methods was made, and there was still no large-scale industry adoption. Attempts were made, notably by Richard Darton at the University of Oxford, to develop methods of process intensification (see Chapter 9) and describe the denaturation of proteins during foam fractionation (see Section 2.5). It was said that Darton was attempting to use foam fractionation to remove the hormone estrogen from the water in the River Thames in the United Kingdom to prevent the fish therein from becoming hermaphroditic, but nothing was published in an archival source to this effect.

The present authors were introduced to foam fractionation in very different ways. Paul Stevenson was nominally working on the process of froth flotation in the heart of Australian coal mining territory when he started to work on a description of the hydrodynamics of dynamic foams in an attempt to model flotation. These descriptions were not universally welcomed by the flotation community, who made the point that they were relevant for demineralized (i.e., gas–liquid) systems whereas there were particles present in flotation columns; the validity of such objections is arguable. Stevenson was looking for a gas–liquid system to apply his work to when Dr. Peter Martin, then of the University of Oxford, suggested the application to the field of foam fractionation. Xueliang Li's approach to foam fractionation was due to his studies in Biochemical Engineering at Hebei University of Technology, China; he still has a collaborative relationship with the Tianjin Kangyi Biotechnology Company, which manufactures the food preservative nisin by, in part, employing foam fractionation, which we believe to be the biggest installation of the process in the world (see Chapter 10). Li modified the plant into its current form and operated it for 2 years, before commencing doctoral studies with Stevenson at the University of Newcastle, Australia. They have concentrated their effort on describing the transport phenomena that underpin foam fractionation, and much of the present volume is based upon Li's (2012) doctoral thesis.

Because foam fractionation is so closely related to froth flotation, and because there has been significantly more research and development effort expended on flotation, it is surprising to the authors that there has been so little cross-fertilization of ideas between the flotation and fractionation communities. It is a recurring theme in this book that concepts developed for froth flotation will be adopted to gain insight into foam fractionation.

The authors would like to think that good progress has been made in understanding the underlying physical phenomena in recent years, and this book is largely an expression of this. It is suggested that true practical advances and industrial adoption will be engendered when innovative engineers use knowledge of the underlying principles to exploit commercial opportunities.

1.5 Applications of Foam Fractionation

As discussed in Section 1.1, by far the most common application of foam fractionation is the protein skimmer found in aquaria, and it is anticipated that this book will have some utility to designers of such devices. However, the potential of foam fractionation, and its allied techniques of ion flotation and bubble fractionation, extends far beyond the fish tank. There have been a myriad of studies on the use of foam fractionation in various systems that have almost entirely been on the laboratory scale and have tended to fall into three categories:

1. The investigation of systems stabilized by organic surfactants to demonstrate one feature or another of the general foam fractionation process
2. The application of foam fractionation to process solutions containing amphipathic biological species such as enzymes or proteins
3. The recovery of metal ions in the process commonly known as "ion flotation"

It is not the intention of this section to provide a comprehensive survey of every study that has been undertaken on foam fractionation, but rather to discuss work that illuminates potential commercial opportunities. Certainly, it is fair to say that it is established that foam fractionation can remove amphipathic species from solution, and the many studies published each year that conclude only this have ever-diminishing value and need not be considered herein.

A comprehensive review of systems to which foam fractionation has been applied has recently been undertaken by Burghoff (2012), and recently there has been a surge in application-related research into the process. There appears to be great potential in employing foam fractionation as an upstream separation process in the biotechnology industry. The biomolecules are often produced in dilute solutions, and their separation can be a significant element of production costs, but their functionality is sensitive to extremes of temperature and pH, so relatively benign separation methods are required. At first inspection, foam fractionation seems to be an appropriate technology because it is benign and cheap, and thus can be used upstream of more expensive chromatographic separation processes. Foam fractionation can exhibit high recovery (i.e., proportion of the target molecule removed from solution), but it can also exhibit poor selectivity in that it indiscriminately tends to remove all amphipathic species. Thus, foam fractionation is typically much better suited to a role as a preliminary upstream separation step. It is also stressed that, although often overlooked in the literature, there is a potential limitation to the process in that it can often denature the target molecules as they adsorb to the gas–liquid surface.

Relatively recent examples of the laboratory-scale use of foam fractionation in bio-separations include those for the removal of enzymes (see, for example, Lambert et al., 2003; Gerken et al., 2006; Linke et al., 2007, 2009) and proteins (Lockwood et al., 1997; Brown et al., 1999; Maruyama et al., 2007; Chan et al., 2007). The fact that foam fractionation can be employed to remove enzymes and proteins from solution is certainly well established. However, whether one would seek to employ the method commercially depends on many other factors, and clearly the industry has not, until now, generally considered the process to be viable. We recall the advice that a former colleague of one of the current authors gives to his chemical engineering students at a prominent UK university: "Just because it is thermodynamically possible for me to stick a needle in my eye, it doesn't mean that it is a clever thing for me to do." We are of the opinion that foam fractionation in bio-processes is potentially a clever thing to do, but industrial decision makers will be persuaded of this by rigorous design and costing information, and not by yet another study that shows that foam fractionation is thermodynamically possible.

There is a notable application for foam fractionation as the stripper of contaminants in wastewater that appears to have achieved some industrial adoption. The Australian state research organization CSIRO has commercialized a foam fractionation unit for the removal of contaminants from water produced when gas is extracted from coal seams. There are major environmental concerns about toxic elements in such water, and foam fractionation is suggested in conjunction with other separation processes to remove such contaminants.

Ion flotation is often considered a separate technology to foam fractionation, but we prefer to consider them as essentially the same process, in that species adsorb to bubbles and are removed in the froth. The essential difference is that a "collector" is used, very much in the same way as it is in coal and mineral flotation, to enable the ionic species to adsorb, and that an electrolysis step is required to recover the metal product from the collected foam. According to Doyle (2003), who wrote an excellent review on ion flotation, it has not been adopted in commercial hydrometallurgical operations, despite there being many laboratory demonstrations of its efficacy (for examples, see Galvin et al., 1997; Filippov et al., 1997; Doyle and Liu, 2003). Just as with foam fractionation, the story of ion flotation is one of repeated laboratory demonstration but little commercial uptake.

1.6 Removal of Contaminants in Rivers

Around the year 2000, public health concerns were raised by a general practitioner in the town of St. Helens in Tasmania, Australia, that she was observing abnormally high incidences of certain diseases, including a range of gut cancers. Around the same time, local Pacific oyster farms started observing dramatic mortality rates in the Georges Bay area (at the mouth of the George River), and several years later a devastating disease struck the local population of Tasmanian Devils. Public water supply for the St. Helens community is drawn from the George River. Suspecting that these observations were due to the use of chemicals used in the local forestry industry, the concerned parties ran their own investigation into the quality of water in the George River.

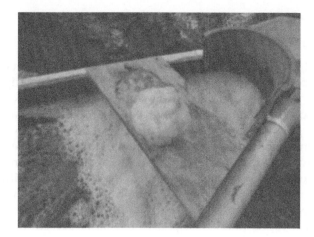

FIGURE 1.4

The "skimmer box" device used in sampling water from the George River in Tasmania. (From Batley, G., Crawford, C., Moore, M., McNeil, J., and Reid, J. 2010. *Report of the George River Water Quality Panel*, Report commissioned by the Government of Tasmania, 29th June 2010, http://www.stors.tas.gov.au/au-7-0150-00001, accessed March 5, 2013.)

Shown in Figure 1.4 is the device called a "skimmer box" that consists of two wooden arms that form the shape of a "V," with the opening facing the upstream direction. It was observed that a large amount of foam collected within the skimmer box and, when sampled and collapsed, proved toxic to cladocerans (water fleas). This troubling finding led to a documentary about the issue that aired on the "Australian Story" program of the ABC and subsequently the Government of Tasmania commissioned an enquiry. One of its findings (Batley et al., 2010; one of the present authors was consulted by the enquiry) was that the foam collected in the skimmer box was being continually enriched in an undefined toxin present in the George River. Thus, it is apparent that a process of foam fractionation was occurring in the skimmer box; and whereas the foamate was toxic to the water fleas, the native river water was not. It was therefore concluded that the water in the George River posed no threat to public health or aquaculture.

While it is true that the skimmer box does indeed appear to be concentrating toxic contaminants via the foam fractionation process, it does not necessarily follow that the lower concentrations of toxins in the main water stream of the George River pose no ecological threat. Clearly, the chemical species that are toxic to water fleas demonstrate some surface activity; otherwise they would not be concentrated in the foam of the skimmer box. Thus, it is quite likely that such toxin will tend to adsorb over time to solid surfaces within the stream, and perhaps to Pacific oysters. It is noted that crushed mussel shells have recently been shown to be effective as a component of fabricated adsorbent (Sakulkhaemaruethai et al., 2010), and it is possible that oyster shells will exhibit similar characteristics.

1.7 Structure of This Book

Nguyen and Schulze (2003) described froth flotation as a microcosm of surface and colloidal science, and this is a fairly succinct description of the process. The physics that underpin flotation involve complex multiphase gas–liquid–solid flows and their interaction with each other at interfaces. It could be argued that the addition of diesel as a collector also temporarily creates an emulsion that may, under certain circumstances, need to be considered in order to fully understand the physical basis of the process. There have been some bold researchers, such as Neethling and Cilliers (2003), who have claimed to have produced complete mechanistic models of the flotation process.

With respect to the number of phases of matter present in the system, froth flotation with potentially four phases present would appear to be significantly more complicated than foam fractionation that has only two phases. However, foam fractionation works by amphipathic species adsorbing to the surface of bubbles. This adsorbed layer can, arguably, be considered a third phase, and this phase is complex indeed. When macromolecules adsorb to a gas–liquid interface, they can change conformational state, and this potentially has deleterious consequences for a foam fractionation column designed to, for instance, enrich solutions of proteins for pharmaceutical applications. Thus, we can see that foam fractionation is a very close cousin of froth flotation, and the lack of a solid phase does not make it any less complex.

Nguyen and Schulze's (2003) approach to the description of flotation was to consider it an amalgamation of colloidal systems; they first considered the nature of dynamic gas–liquid foam and then considered particulate phenomena, such as attachment of hydrophobic particles to bubbles (which is analogous to molecular adsorption in foam fractionation). We take a similar approach in this book. We begin the substantive portion on the book by considering molecular adsorption to gas–liquid interfaces, with special emphasis on situations relevant to foam fractionation. Thus, difficulties introduced due to the convection of liquid past a surface *vis-a-vis* adsorption and changes in conformational state of macromolecules will be discussed, although it is immediately conceded that there is incomplete knowledge of both phenomena.

The behavior of foam fractionation devices is, of course, critically dependent on the nature of vertical dynamic columns of foam. We believe that significant advances have been made in the past decade on mechanistically understanding such systems; this work will be described in the context of earlier treatments. In addition, the fluid mechanics of bubble swarms will be investigated in Chapter 5 as much adsorption is thought to occur in the bubbly liquid, and this zone contributes significantly to the liquid inventory in a foam fractionation column.

Once a treatment of the fundamental processes of the adsorption of macromolecules and the fluid mechanics of foam and bubbly liquids has been presented, these eclectic concepts can be employed in discussions about process design of foam fractionation. The first decision for the designer to take is whether the process should be run continuously or batchwise, and whether it should be operated with an external reflux stream. The relevant relative advantages and disadvantages are discussed. To make foam fractionation work, bubbles must be created and then, after the foam has been collected, the bubbles must collapse. All these processes must be designed, and therefore they are discussed herein. Design methods for the physical dimensions of plain columns are described, before novel designs to facilitate process intensification are discussed. The book concludes, in Chapter 10, with a case study of the manufacture of nisin, which we believe to be currently the largest foam fractionation installation in the world.

Foam fractionation is a technology that is very much in its infancy, and there are no definitive design codes. It is hoped that by presenting discussions of the physics of the process alongside practical considerations, the engineer will be guided to develop designs appropriate to their specific duty. We hope to make a case for a logical design approach to foam fractionation rather than a trial-and-error method which we think has been a contributor to the low industrial adoption of the process, although some of the latter strategy is unavoidable.

Nomenclature

C_b	Concentration in the bulk liquid	[mol.m^{-3}]
C_f	Foamate concentration	[mol.m^{-3}]
E	Enrichment factor	[–]
r_b	Bubble radius	[m]
r_{32}	Sauter mean bubble radius	[m]
S	Specific surface area	[m^{-1}]
ε	Volumetric liquid fraction	[–]
Γ	Surface excess	[mol.m^{-2}]

References

Batley, G., Crawford, C., Moore, M., McNeil, J., and Reid, J. 2010. *Report of the George River Water Quality Panel*, Report commissioned by the Government of Tasmania, 29th June 2010, http://www.stors.tas.gov.au/au-7-0150-00001, accessed March 5, 2013.

Brown, A.K., Kaul, A., and Varley, J. 1999. Continuous foaming for protein recovery. Part I. Recovery of β-casein, *Biotechnology and Bioengineering*, 62, 278–279.

Burghoff, B. 2012. Foam fractionation applications, *Journal of Biotechnology*, 161, 126–137.

Cengel, J.D., and Lemlich, R. 1989. A technique for studying the aging of liquid foam under simulated conditions of microgravity, *Journal of Colloid & Interface Science*, 128, 608–610.

Chan, N.Y., Hossain, M.M., and Brooks, M.S. 2007. A preliminary study of protein recovery from mussel blanching water by a foaming process, *Chemical Engineering and Processing*, 46, 501–504.

Charm, S.E. 1972. Foam separation of enzymes and other proteins, in *Adsorptive Bubble Separation Techniques*, Editor R. Lemlich, Academic Press, Library of Congress Number 75-154398.

Charpentier, J.C. 1981. Mass transfer in gas liquid absorbers and reactors, *Advances in Chemical Engineering*, 11, 3–133.

Couper, J.R., Penney, W.R., Fair, J.R., and Walas, S.M. 2010. Gas-solid separations, in *Chemical Process Equipment*, 2010, Gulf Professional Publishing, Boston.

de Vries, A.J., 1972. Morphology, Coalescence and size distribution of foam bubbles, in *Adsorptive Bubble Separation Techniques*, Editor R. Lemlich, Academic Press, Library of Congress Number 75-154398.

Doyle, F.M. 2003. Ion flotation-Its potential for hydrometallurgical operations, *International Journal of Mineral Processing*, 72, 387–399.

Doyle, F.M., and Liu, Z. 2003. The effect of triethylenetetraamine (Trien) on the ion flotation of Cu^{2+} and Ni^{2+}, *Journal of Colloid & Interface Science*, 258, 396–403.

Escobal, P.R. 2000. *Aquatic Systems Engineering: Devices and How They Function*, Dimension Engineering Press, ISBN 1888381108.

Filippov, L.O., Houot, R., and Joussemet, R. 1997. Physicochemical mechanisms and ion flotation possibilities using columns for Cr^{6+} recovery from sulphuric solutions, *International Journal of Mineral Processing*, 51, 229–239.

Finch, J.A., and Dobby, G.S. 1990. *Column Flotation*, Pergamon Press, ISBN 0080401864.

Fuerstenau, M.C., Jameson, G.J., and Yoon, R.-H. 2007. Froth flotation: A century of innovation, in *Society for Mining Metallurgy & Exploration*, ISBN 0873352521.

Galvin, K.P., Nicol, S.K., and Waters, A.G. 1992. Selective ion flotation of gold, *Colloids & Surfaces*, 64, 21–23.

Gerken, B.M., Nicolai, A., Linke, D., Zorn, H., Berger, R.G., and Parlar, H. 2006 Effective enrichment and recovery of laccase C using continuous foam fractionation, *Separation & Purification Technology*, 49, 291–294.

Gol'dfarb, I.I., Kann, K.B., and Schreiber, I.R. 1988 Liquid flow in foams, *Izv Akad Nauk*, 23, 244–249.

Gross-Sorokin, M.Y., Roast, S.D., and Brighty, G.C. 2006. Assessment of feminization of male fish in English rivers by the Environment Agency of England and Wales, *Environmental Health Perspectives*, 114 (S-1), 147–151.

Johnson, A., and Darton, R. 2004. Removing oestrogenic compounds from sewage effluent, http://www.eng.ox.ac.uk/chemeng/people/darton/oestrogens.pdf, accessed March 5, 2013.

Lambert, W.D., Du, L., Ma, Y., Loha, V., Burapatana, V., Prokop, A., Tanner, R.D., and Pamment, N.B. 2003. The effect of pH on the foam fractionation of β-glucosidase and cellulose, *Bioresource Technology*, 87, 247–253.

Lemlich, R. 1972. *Adsorptive Bubble Separation Techniques,* Academic Press, Library of Congress Number 75-154398.

Lemlich, R., and Lavi, E. 1961. Foam fractionation with reflux *Science,* 134, 191.

Leonard, R.A., and Lemlich, R. 1965a. Laminar longitudinal flow between close-packed cylinders, *Chemical Engineering Science,* 20, 790–791.

Leonard, R.A., and Lemlich, R. 1965b. A study of liquid flow in foam I. Theoretical model and application to foam fractionation, *AIChE Journal,* 11, 18–25.

Leonard, R.A., and Lemlich, R. 1965c. A study of liquid flow in foam II. Experimental verification and observations, *AIChE Journal,* 11, 25–29.

Li, X. 2012. Transport Phenomena in Foam Fractionation, PhD thesis, The University of Newcastle (Australia).

Linke, D., Zorn, H., Gerken, B., Parlar, H., and Berger, R.G. 2007. Laccase isolation by foam fractionation – New prospects of an old process, *Enzyme and Microbial Technology,* 40, 273–277.

Linke, D., Nimtz, M., Berger, R.G., and Zorn, H. 2009. Separation of extracellular esterases from pellet cultures of the basidiomycete Pleurotus sapidus by foam fractionation, *Journal of the American Oil Chemists' Society,* 86, 437–444.

Lockwood, C.E., Bummer, P.M., and Jay, M. 1997. Purification of proteins using foam fractionation, *Pharmaceutical Research,* 14, 1511–1515.

Maruyama, H., Seki, H., Suzuki, A., and Norio, I. 2007. Batch foam separation of a soluble protein. *Water Research,* 41, 710–718.

Neethling, S.J., and Cilliers, J.J. 2003. Modelling flotation froths, *International Journal of Mineral Processing,* 72, 267–287.

Nguyen, A.V., and Schulze, H.J. 2003. *Colloidal Science of Flotation,* Marcel Dekker, ISBN 0824747828.

Prokop, A., and Tanner, R.D. 1993. Foam fractionation of proteins: Potential for separations from dilute starch suspensions, *Starch-Stärke,* 45, 150–154.

Saint-Jalmes, A., Marze, S., Safouane, M., and Langevin, D. 2006. Foam experiments in parabolic flights: Development of an ISS facility and capillary drainage experiments, *Microgravity Science & Technology,* 18, 22–30.

Sakulkhaemaruethai, S., Duangduen, C., Pivsa-Art, W., and Pivsa-Art, S. 2010. Fabrication of composite material from sea mussel shells and white clay as a versatile sorbent, *Energy Research Journal,* 1, 78–8.

Stevenson, P. 2012. Gas-liquid mass transfer in foam, in *Foam Engineering: Fundamentals & Applications*, Editor P. Stevenson, Wiley, ISBN: 0470660805.

Verbist, G., Weaire, D., and Kraynick, A.M. 1996. The foam drainage equation, *Journal of Physics: Condensed Matter,* 8, 3715–3731.

Wenzel, R.N. 1949. Surface roughness and contact angle, *Journal of Physical and Colloid Chemistry,* 53, 1466–1467.

Yang, C., and Gu, Y. 2004. Modeling of the adsorption kinetics of surfactants at the liquid-fluid interface of a pendant drop, *Langmuir,* 20, 2503–2511.

2

Adsorption of Surface-Active Species to Gas–Liquid Interfaces

2.1 The Reduction in Gibbs Free Energy upon Adsorption

The very reason that foam fractionation is effective as a process is that certain molecular species want to adsorb to gas–liquid interfaces, just as hydrophobic particles attach to the surface of bubbles in flotation. Thus, the process of adsorption of molecular species onto gas–liquid interfaces is fundamental to foam fractionation, and this will be the exclusive topic of Chapter 2. Topics to be addressed will be the equilibrium and rate of surface adsorption for various species, as well as how these properties change when liquid advects past a bubble surface, as it does in foam fractionation. The difference between an *equilibrium* adsorption condition and a *rate* of adsorption is often conflated, and is the essential difference between the material that typically interests chemists versus that of interest to chemical engineers.

However, for any physical process to proceed, including those in mass transfer, a *driving force* is required. If it is energetically more favorable for certain molecular species to reside at the gas–liquid surface, then molecules will diffuse to the bubble surface where they adsorb. The adsorption of such a molecule to the surface will result in a reduction in the surface tension (which is synonymous with the surface energy). To express this in another way, one can say that the Gibbs free energy of the system has decreased, and it is this potential for energy reduction that causes adsorption; the bigger the potential for reduction of energy, the greater the rate of adsorption. Eventually, a surface concentration (commonly known as a "surface excess") is attained such that any further net adsorption will not further reduce the Gibbs free energy, and the system is said to be at equilibrium. This is an equilibrium relationship between the bulk concentration of species (i.e., the concentration within the liquid and has units of mol.m^{-3}) and the surface excess (which has units of mol.m^{-2}), takes the name "adsorption isotherm," and it is the nature of such relationships that is discussed in Section 2.2.

2.2 Equilibrium between Bulk Concentration and Surface Excess

Consider Figure 2.1, which shows a general surface-active (surfactant) species that is dissolved in the bulk liquid. The concentration of this species is C_b, and this is in equilibrium with the surface concentration, or *surface excess*, denoted by the symbol Γ. The equilibrium relationship between concentration in the bulk liquid and the surface excess is known as the *adsorption isotherm*. Direct measurements of the surface excess, although possible, are generally difficult to perform, but the surface excess can be conveniently estimated indirectly by measurements of the equilibrium surface tension, σ, via Gibbs isotherm, which states that

$$\Gamma = -\frac{C_b}{nRT}\frac{d\sigma}{dC_b} \tag{2.1}$$

where T is the absolute temperature and R is the universal gas constant; n is a number that, for the present purposes, takes the value of 1 for nonionic surfactants and 2 for completely disassociated ionic surfactants. Great care should be taken when measuring the surface tension, for it is specifically the *equilibrium surface tension* that is required, as surface-active species adsorb to a surface over time; the *dynamic surface tension* asymptotically approaches an equilibrium value, which may be conveniently estimated from dynamic

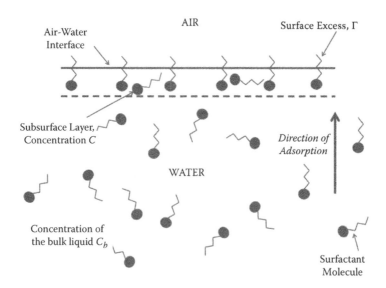

FIGURE 2.1
A schematic representation of surfactant molecules at the water–air interface.

data using the method of Makievski et al. (1997). Appropriate techniques for measuring dynamic surface tension include the pendant bubble and droplet methods that are critiqued by Li et al. (2010). Figure 2.2 shows dynamic surface tension data for the protein bovine serum albumin measured using the pendant bubble method, and Figure 2.3 extrapolates the same data to infinite

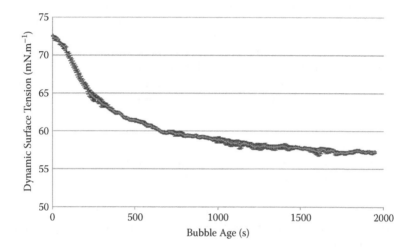

FIGURE 2.2
The evolution of dynamic surface tension of 0.7 g l⁻¹ bovine serum albumin solution at 22°C measured using the pendant bubble method. The error bars represent the standard error about the mean of five trials. (Data courtesy of Claire Burnett.)

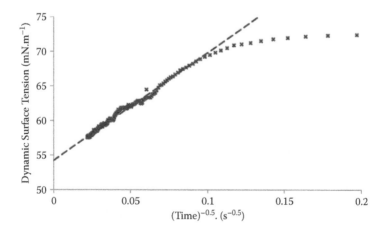

FIGURE 2.3
Dynamic surface tension data of bovine serum albumin solution at a concentration of 0.7 g l⁻¹ and at 22°C, replotted using the method of Makievski et al. (1997) to reveal an equilibrium surface tension of 54.2 mN m⁻¹. (Data provided courtesy of Claire Burnett.)

time using the Makievski method to reveal an equilibrium surface tension of 54.2 mN.m^{-1}. Techniques that measure the *static surface tension* of a chemically pure liquid, such as the Wilhelmy Plate method, are inappropriate because they can only measure a single time-invariant value of the surface tension.

An adsorption isotherm for the protein bovine serum albumin at a water-air interface at 22°C is given in Figure 2.4; all data were collected by measuring the dynamic surface tension using the pendant bubble method and extrapolated to long times using the Makievski method. It should be noted that the bulk concentration is quoted in units of moles per cubic meter rather than the more commonly encountered concentration unit of moles per liter. This is because the adsorption isotherm relates bulk concentration with surface excess, which has units of moles per square meter, and a consistent unit system must be adopted if a mathematical approximation to adsorption isotherm data is to be proposed.

To those uninitiated in surface science, it may already appear that there are a lot of "adsorption isotherms" to consider, and the term certainly confused the present authors when they first ventured into this field. For the present purposes, it may be helpful to offer the following simple definitions:

1. *The Adsorption Isotherm:* The equilibrium relationship between bulk concentration and surface excess, which, in practice, must be found experimentally.

2. *Gibbs Adsorption Isotherm:* The relationship enabling the equilibrium surface excess to be inferred from data for the equilibrium surface tension versus bulk concentration, given in Equation 2.1.

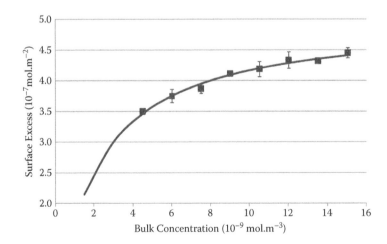

FIGURE 2.4
Adsorption isotherm data for bovine serum albumin at 22°C, with the Langmuir isotherm given in Equation 2.7 superimposed. Error bars represent the standard error on data gained by the dynamic surface tension method. (Data provided courtesy of Claire Burnett.)

3. *The Langmuir Adsorption:* Just one of many mathematical functions, some of which are described below, that are a mathematical fit to experimental adsorption isotherm data and serve as a convenient representation of data in many circumstances. Some common adsorption isotherms, as given in the excellent review of Eastoe and Dalton (2000), are as follows:

a. *Henry's Law Isotherm.* This is the simplest expression of an adsorption isotherm, and is simply a linear relationship between surface excess and bulk concentration. Many systems exhibit such a relationship at low concentrations; for example, the Langmuir adsorption isotherm, described below, also exhibits such a characteristic:

$$\Gamma = K_H C_b \tag{2.2}$$

$$C_b = \Gamma / K_H \tag{2.3}$$

where K_H is the surface Henry's Law constant. Often, a *surface equation of state* may be quoted for an adsorption isotherm, which relates the surface pressure Π (i.e., the difference between the surface tension of the fresh surface that is devoid of adsorbing species and the equilibrium surface tension) with the surface excess and absolute temperature. The surface equation of state can simply be derived by substituting the mathematical function for the adsorption isotherm into the Gibbs isotherm (Equation 2.1). The surface equation of state corresponding to the Henry's Law isotherm is

$$\Pi = \sigma_0 - \sigma = nRT\Gamma \tag{2.4}$$

b. *Langmuir Isotherm.* The Langmuir isotherm is commonly adopted in studies of foam fractionation because it is relatively simple, but approximately captures the nature of the adsorption isotherm at the relatively high surface excesses that are desirable in the process.

$$\Gamma = \Gamma_\infty \left(\frac{K_L C_b}{1 + K_L C_b} \right) \tag{2.5}$$

or, making the bulk concentration the subject of the expression,

$$C_b = \frac{1}{K_L} \frac{\Gamma}{\Gamma_\infty - \Gamma} \tag{2.6}$$

where K_L is a constant and Γ_∞ is the saturation surface excess. The latter term warrants further explanation: It is the maximum value of surface excess that can be supported by a gas–liquid interface, and is a property that has profound consequences in foam fractionation practice. An increase in bulk concentration beyond that which approximately corresponds to the saturation surface excess will not engender any further adsorption; and because foam fractionation works by species adsorbing to surfaces, it can be seen that there is no substantive advantage gained from operating above this bulk concentration. The Langmuir isotherm predicts a surface excess that is asymptotic to the saturation surface excess.

Superimposed onto Figure 2.4 is a Langmuir isotherm fitted to the experimental data for BSA, which is given by

$$\Gamma = 5 \times 10^{-7} \left(\frac{5 \times 10^8 C_b}{1 + 5 \times 10^8 C_b} \right) \tag{2.7}$$

which implies a saturation surface excess of 5×10^{-7} mol.m^{-2}; the value of K_L is 5×10^8 m^2.mol^{-1}.

The corresponding surface equation of state is

$$\Pi = \sigma_0 - \sigma = -nRT\Gamma_\infty \ln\left(1 - \frac{\Gamma}{\Gamma_\infty} \right) \tag{2.8}$$

It is noted that, as the bulk concentration becomes vanishingly small, a linear Henry's Law-type isotherm is recovered from the Langmuir isotherm, that is,

$$\Gamma = K_L \Gamma_\infty C_b \tag{2.9}$$

c. *Frumkin Isotherm*

$$C_b = \frac{1}{K_F} \frac{\Gamma}{\Gamma_\infty - \Gamma} \exp\left[-A\left(\frac{\Gamma}{\Gamma_\infty} \right) \right] \tag{2.10}$$

where K_F is a constant and A is called the surface interaction parameter. The corresponding surface equation of state is

$$\Pi = \sigma_0 - \sigma = -nRT\Gamma_\infty \ln\left(1 - \frac{\Gamma}{\Gamma_\infty} \right) - \frac{nRTA}{2}\Gamma_\infty \left(\frac{\Gamma}{\Gamma_\infty} \right)^2 \tag{2.11}$$

d. *Freundlich Isotherm*

$$\Gamma = kC_b^{1/N} \tag{2.12}$$

$$C_b = \left(\frac{\Gamma}{k}\right)^N \tag{2.13}$$

where k is a constant and N is a consistency index. The corresponding surface equation of state is

$$\Pi = \sigma_0 - \sigma = NnRT\Gamma \tag{2.14}$$

e. *Volmer Isotherm*

$$C_b = K_v \left(\frac{\Gamma}{\Gamma_\infty - \Gamma}\right) \exp\left[\frac{\Gamma}{\Gamma_\infty - \Gamma}\right] \tag{2.15}$$

where K_v is a constant with the units of bulk concentration. The corresponding surface equation of state is

$$\Pi = \sigma_0 - \sigma = \frac{\Gamma_\infty^2}{\Gamma_\infty - \Gamma} nRT \tag{2.16}$$

The above five mathematical descriptions to adsorption isotherms are just a selection of those found in the surface science literature. However, in studies of foam fractionation, it is common to see only the Langmuir Isotherm because of its simplicity and recognition of the existence of a saturation surface excess.

2.3 Diffusion-Controlled Adsorption of Molecules to Gas–Liquid Surfaces

One of the main distinctions between chemists and chemical engineers is that, whereas the former are typically interested in equilibrium conditions (as discussed in the previous section in the context of species adsorbing to gas–liquid interfaces), the latter are often more interested in the rate at which processes occur.

The driving force for the adsorption of a surfactant molecule is the potential for a reduction in Gibbs free energy, as discussed in Section 2.1.

Before beginning a mechanistic description of the process of molecular adsorption to a gas–liquid interface, it is necessary to define one more concentration: the *subsurface concentration*. This can be considered the concentration of the adsorbing species in the liquid directly in contact with the gas–liquid interface, and the difference in concentration between the bulk and subsurface concentrations provides the driving force for molecular diffusion toward the interface. Surfactant adsorption to a gas–liquid interface is shown schematically in Figure 2.4.

The adsorption of a nonionic surfactant molecule onto an interface is generally considered to be governed by two processes: (1) the diffusion of the surfactant molecule to the subsurface from the bulk solution; and (2) the adsorption of the molecule from the subsurface onto the interface, which can be considered a rearrangement step. "Diffusion-controlled" models assume an immediate local equilibrium between the subsurface and the interface; thus, the rate of adsorption is determined by the mass transfer from the bulk solution to the subsurface, which is described by Fick's Law of diffusion. The diffusion equations can be solved by applying numerical methods when proper initial conditions and boundary conditions are provided. However, an analytical expression is desirable to gain a deeper understanding of the adsorption process. The diffusion equation was first integrated by Ward and Tordai (1946), resulting in an integral equation of the form

$$\Gamma(t) = 2\sqrt{\frac{D}{\pi}} \left\{ C_b \sqrt{t} - \int_0^{\sqrt{t}} C(\tau) d\left(\sqrt{t - \tau}\right) \right\} \tag{2.17}$$

where t is time since the formation of the fresh surface, D is the coefficient of molecular diffusion, and τ is a dummy variable with the units of time. C_b and $C(\tau)$ are the bulk concentration and the subsurface concentration, respectively. The second term within the bracket is the so-called back-diffusion term, and it is similar in form to the convolution integral.

The application of Equation 2.17 to estimate the evolution of surface excess on a gas–liquid interface is problematic because it requires information about the subsurface concentration, so it is necessary to assume a mathematical description of the adsorption isotherm to enable integration. Only if the simplest description of the adsorption isotherm, the Henry's Law isotherm, is adopted can Equation 2.17 be analytically integrated. In fact, in almost all circumstances, Equation 2.17 must be numerically integrated, and an example of the implementation of a robust numerical scheme is that of Li et al. (2010). It should be noted that the Ward-Tordai procedure is for a planar gas–liquid surface, and that adsorption onto curved surfaces is a function of the radius of curvature. The adsorption rate to a surface that is convex with respect to the gas phase (i.e., a bubble) is enhanced, with the effect becoming greater as the bubble becomes smaller. A modification of the Ward-Tordai description for such circumstances was first given by Mysels (1982):

$$\Gamma(t) = 2\sqrt{\frac{D}{\pi}}\left\{C_b\sqrt{t} - \int_0^t C(\tau)\,\mathrm{d}\sqrt{t-\tau}\right\} + \frac{D}{r}\left\{C_b t - \int_0^t C(\tau)\,\mathrm{d}(t-\tau)\right\} \quad (2.18)$$

However, for the size of bubbles typically encountered in foam fractionation, the adsorption rate enhancement due to a curved bubble surface is not substantive, and for the present purposes the effect will be disregarded.

The time taken for the surface excess to approach the asymptotic equilibrium value depends on several factors, not the least of which is the coefficient of molecular diffusion. However, relatively small molecules with high coefficients of molecular diffusion can exhibit an asymptotic surface excess on the order of several seconds, whereas larger molecules may take several hours to approach the equilibrium condition, although such molecules are more likely to require a rearrangement step to adsorb as is described below.

It is important to note that the Ward-Tordai equation is valid only for the diffusion of nonionic species. The rate of adsorption of ionic surfactants is much more complicated due to the presence of the electric double layer; see Section 2.6. However, this limitation of validity is sometimes overlooked. For example, Phan et al. (2005) and Phan et al. (2012) tried to apply it to the adsorption of sodium dodecyl benzene sulfonate (SDBS) and cetyl trimethyl-ammonium bromide (CTAB), respectively, to gas–water interfaces.

In order to adopt the Ward-Tordai equation, one requires knowledge of the coefficient of molecular diffusion. This can be estimated by the well-known correlation of Wilke and Chang (1955) or the correlation of Young et al. (2004) that was specifically developed for proteins. Methods of measuring the coefficient of molecular diffusion are explained by Cussler (2009). Measurement techniques include the nuclear magnetic resonance method of Stejskal and Tanner (1965).

2.4 Adsorption of Macromolecules

The model of diffusion-controlled adsorption assumes that the subsurface layer concentration is instantaneously in equilibrium with the surface excess; the kinetics of adsorption are governed by the diffusion of the adsorbing species from the bulk liquid to the subsurface, and nothing else. As soon as the adsorbing species reaches the subsurface through diffusion, it adsorbs to the surface and there is no further kinetic step. However, there is a body of experimental data that shows that, for some systems, this is an incomplete description, and that there is an activation energy required for adsorption once the molecule has reached the subsurface. In general, this energy barrier is more significant for large molecules (i.e., *macromolecules*) than it is for smaller, simpler surfactants.

A mixed diffusion-kinetic model was suggested by Miller and Kretzschmar (1980), a solution to which was given by Borwankar and Wasan (1983), the details of which are too convoluted to discuss herein. It is worth stating that the existence of an energy barrier at the surface reduces the rate of adsorption from the prediction of diffusion-controlled adsorption.

The adsorption of macromolecules, such as proteins, is complicated still further. Quite apart from the crucial consideration of protein unfolding upon adsorption (discussed in Section 2.5), rather than species adsorbing in a single monolayer at the bubble surface, multilayer adsorption of protein molecules is also possible. Although experimental studies have been conducted (Ybert and di Meglio, 1998; Miller et al., 1998; Makievski et al., 1999), quantitative modeling of the dynamic adsorption of proteins remains in its infancy.

In addition to the considerations above, it is apparent that the pH of a solution has an important role in determining the adsorption rate and extent of protein adsorption to gas–liquid interfaces. The pH at which the protein exhibits zero net charge is known as the isoelectric point. The data of Brown et al. (1999) and Du et al. (2003), as well as unpublished data of our own, suggest that operation at the isoelectric point is desirable *vis-a-vis* adsorption rate and extent. The isoelectric point of a protein molecule can be determined by analytical methods, including that of Sillero and Maldonado (2006).

2.5 Protein Denaturation at Gas–Liquid Interfaces

Protein molecules, in their native conformation, are interconnected polypeptides arranged in specific structures. Generally speaking, both hydrophobic and hydrophilic groups exist at the surface of a protein molecule. At the air-water interface, these hydrophobic groups tend to escape from the water side to spread at the interface, while the hydrophilic groups prefer to stay in the water side. Given enough time, the tertiary conformation of the protein molecule will be changed, sometimes combined with the loss of its bioactivity if the change is irreversible (see, for instance, Miller et al., 1998). Such loss of bioactivity is commonly known as *denaturation* and can have a deleterious impact upon the functionality of the molecule because denaturation tends to be irreversible. In the industrial foam fractionation of nisin, a 10% denaturation is typical (see Chapter 10). It can be seen that, in the case of the foam fractionation of proteins for example, the very method of recovering the molecules from solution can preclude their efficient subsequent use, but this tends to be overlooked. Thus, it is very important to discuss denaturation herein.

Graham and Philips (1979a, b) published an extensive study on the kinetics of the adsorption and denaturation of proteins at gas–liquid interfaces. They showed (1979a) that, when the protein bovine serum albumin adsorbed, the surface pressure (i.e., the difference in the initial surface tension and the dynamic surface tension) changed simultaneously with the surface excess. However, when the globular protein lysozyme adsorbed to the gas–liquid interface, the surface pressure increased at a greatly longer time constant than the surface excess, indicating significant unfolding of protein after the surface excess had approached its equilibrium value. In addition, it was inferred that protein adsorption was diffusion controlled at early stages, but that an energy barrier for adsorption had to be overcome as the surface excess increased. Graham and Philips (1979b) showed that proteins adsorbed in multilayers as described in Section 2.4, but that adsorption to anything but the first layer had no effect on the surface tension.

Twenty years after the work of Graham and Philips, there were several studies that investigated denaturation specifically in the foam fractionation process, but the implications of this body of work are often overlooked by workers aiming to show that foam fractionation has utility for the removal of proteins from solution. Liu et al. (1998) demonstrated the denaturation of enzymes during foam fractionation, and denaturation of the protein β-casein was detected during foam fractionation using a variety of experimental techniques, including circular dichroisim. Clarkson et al. (1999a) showed that some protein denaturation was reversible (i.e., some molecules returned to their native conformational state upon desorption from the gas–liquid interface); significantly more β-casein molecules returned to their native state upon desorption than did molecules of the enzyme pepsin. In addition, it was concluded that protein denaturation was because of conformational changes and that damage due to oxidation was negligible. Clarkson et al. (1999b) detected changes in conformational state in other proteins. Thus, it is clear that the foam fractionation of proteins and enzymes causes some denaturation, but the degree of denaturation depends on the particular molecule that is adsorbing.

Burnett and Stevenson (2012) suggested a novel and simple method of qualitatively assessing the degree of denaturation due to repeated cycles of surface adsorption and desorption. They repeatedly foamed aqueous solutions of bovine serum albumin and then allowed the foam to collapse before measuring the dynamic surface tension. The surface tension evolution was then compared to that of a control sample that had not been foamed using the pendant bubble method. It can be seen (see Figure 2.5) that the rate of decrease in surface tension was greater for the pre-foamed solution than for the control, and the equilibrium surface tension that was approached was lower, clearly indicating that irreversible protein denaturation occurred upon adsorption to gas–liquid interfaces.

It has been suggested by us that foam fractionation might find utility in the removal of proteins from whey or milk emulsion. When air bubbles

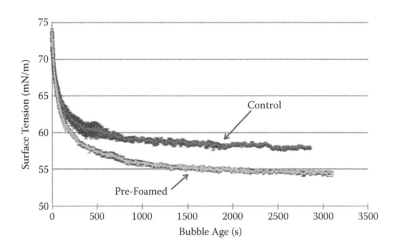

FIGURE 2.5
The dynamic surface tension of 0.5 g/l BSA solution at 22°C, showing that the pre-foamed sample exhibits greater denaturation than the control sample. (Adapted from Burnett, C.R., and Stevenson, P. 2012. Denaturation of proteins upon adsorption to bubble surfaces, Paper 2, *CHEMECA '12*, Wellington, New Zealand, 23–26 September.)

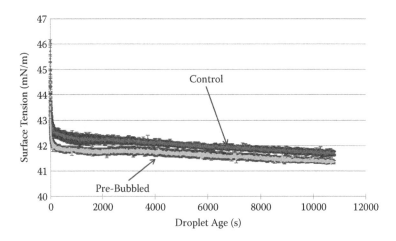

FIGURE 2.6
The dynamic surface tension of full fat milk at 22°C, showing that the pre-bubbled and control samples do not exhibit significantly different surface tension evolution behaviors. (Adapted from Burnett, and C.R., Stevenson, P. 2012. Denaturation of proteins upon adsorption to bubble surfaces, Paper 2, *CHEMECA '12*, Wellington, New Zealand, 23–26 September.)

are sparged to milk, a stable foam is not produced, and therefore any foam fractionation technology would require the use of a co-surfactant to help stabilize the froth. However, by bubbling air bubbles through milk and comparing the dynamic surface tension to that of a control (i.e., nonbubbled, see Figure 2.6), the degree of protein denaturation could be qualitatively assessed. Milk proteins apparently do not denature significantly upon adsorption to gas–liquid interfaces. We can only speculate that the lipids in milk may fulfill some protective role against conformational changes.

2.6 Adsorption of Ionic Species

The complexity that arises in ionic surfactant adsorption, as opposed to nonionic surfactants, is the effect of electrostatic interactions on both the equilibrium state and the rate of transport to the interface, due to the existence of the electric double layer (EDL) near the interface. Figure 2.7 illustrates the electric double layer around a gas bubble.

Generally speaking, the electrostatic repulsion between adsorbed ions decreases the equilibrium surface excess; therefore the adsorbed ions are not closely packed at the interface. Under nonequilibrium conditions, the electrostatic repulsion between the ions inside and outside the adsorption layer retards the adsorption kinetics. Oppositely charged ions (counterions) can be immobilized at the adsorption layer, which forms the Stern layer. Conditions such as pH and ionic strength, which can affect the electrostatic repulsion, will also affect the adsorption kinetics. Quantitative modeling of the adsorption of ionic surfactants emerged in the 1980s (see Dukhin et al., 1983; Miller et al., 1985). The recent review of Karakashev et al. (2008) of the equilibrium adsorption of both ionic and nonionic surfactant at the gas–liquid interface compared and validated various models.

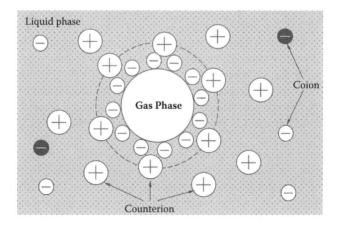

FIGURE 2.7
Illustrative diagram of the electric double layer around a gas bubble in an aqueous solution of ionic surfactant. The dashed circle denotes the Stern layer where counterions are adsorbed.

2.7 Adsorption Modification Due to Liquid Advection Past a Surface

In a typical foam fractionation column, there is a zone of bubbly liquid and a zone of pneumatic foam. There is a net slip downward of the liquid relative to the gas phase. Typically, the slip velocity in the bubbly liquid is significantly greater than in the foam layer. The question arises as to whether or not the existence of a slip velocity modifies the adsorption isotherm of, and the adsorption rate to, the gas–liquid interface. The answer to this is that the slip velocity does indeed modify adsorption behavior with respect to quiescent surfaces, but that such modification is complicated and not even remotely properly understood; the modification is thought more significant in the bubbly liquid where the slip velocity is greater.

The adsorption of surface-active materials modifies the properties of the bubble surface, and hence the stress state of the surface and the drag force exerted on the surface, which affects the fluid flow around the bubble. The fluid flow, in turn, affects the mass transfer and the surfactant distribution on the surface. Experimental and computational fluid dynamical (CFD) studies (Sharifullin and Luebbert 2001; Zhang et al., 2001; Matsumoto et al., 2006) have shown that the adsorption isotherms of quiescent and moving interfaces (i.e., where a slip velocity exists between the liquid and gas phases, as it does in foam fractionation) are different, because of an interdependency between hydrodynamic conditions and adsorption. This effect is more significant in the liquid pool where the bubbles rise rapidly and the slip velocity is high than in the foam bed where the slip velocity is relatively lower. The dynamic adsorption layer theory and the stagnant cap model were developed to describe the interaction between the adsorption/ desorption of surfactant and flow field around a buoyant bubble (Dukhin et al., 1995, 1998). According to the stagnant cap model, a steady state is reached after a certain time since the formation of the bubble. After the attainment of steady state, the total amount of adsorbed substance does not change, because the adsorption rate on one part of the bubble surface is equal to the desorption rate from another part. The movement of the bubble induces a surface excess gradient, such that at the leading pole of the bubble, the surface concentration is lowest while at the rear stagnation point, it is a maximum.

The dynamic adsorption layer theory is only valid for small Reynolds numbers, while in foam fractionators the mixing caused by the motion of the bubbles invalidates the simplifying assumptions demanded by the theory. Therefore, few in the foam fractionation community have considered the phenomenon (see below for a brief discussion). Another reason that the above-mentioned theoretical approaches have rarely been considered in foam fractionation studies is that, although there is an effect of the dynamic nature of the interface, it is more difficult to control as compared to other design parameters, such as the depth of the liquid pool and the foam bed.

The effect of liquid convection past a surface on the rate of adsorption to the surface is apparently very complicated and poorly understood. Thus, we will instead appeal to established work on how slip velocity influences the dissolution mass transfer from a solid sphere into a liquid, or the absorption of a species from a liquid into a solid sphere. It is a well-known phenomenon that liquid flow past a surface generally enhances the mass transfer rate if the process is diffusion controlled. A simple example is that of stirring a cup of tea to increase the rate of sugar dissolution. It can be shown via dimensional analysis that the mass transfer coefficient from an isolated solid sphere— nondimensionalized as a Sherwood number, Sh,

$$Sh = \frac{k_L d}{D} \tag{2.19}$$

where k_L is the fluid film mass transfer coefficient, d is the diameter of the sphere, and D is the coefficient of molecular diffusion, can be expressed as a function of the Reynolds number, Re:

$$Re = \frac{\rho_f V d}{\mu} \tag{2.20}$$

where V is the absolute slip velocity between the solid and fluid phases, and ρ and μ are the density and dynamic viscosity of the fluid phase, respectively, and the Schmidt number, Sc:

$$Sc = \frac{\mu}{\rho_f D} \tag{2.21}$$

There are analogies for heat transfer, which is a purely diffusive process in this context. For $Re < 1$, Levich (1962) analytically derived

$$Sh = 0.66 Re^{1/3} Sc^{1/3} \tag{2.22}$$

For $Re > 1$, Frössling, cited in Tosun (2002), gave the following empirical expression:

$$Sh = 2 + 0.56 Re^{1/2} Sc^{1/3} \tag{2.23}$$

The above two correlations are for the mass transfer from single spheres. The correlation of Wakao and Kaguei (1982) for the mass transfer coefficient for a closely packed bed of solid spheres is given by

$$Sh = 2 + 1.1 Re^{0.6} Sc^{1/3} \tag{2.24}$$

where it is valid for $Re < 3000$. Equations 2.22, 2.23, and 2.24 all indicate that the mass transfer coefficient increases with increasing slip velocity (i.e., the rate of steady convection) past the surface.

In the absence of any definitive data, it could be suggested that Equation 2.23 is used to calculate a modified mass transfer coefficient for the adsorption of surface-active species to the bubble surface in the bubbly liquid of a foam fractionator and Equation 2.24 used to calculate within the foam. However, in the interest of conservative design, it is arguably more appropriate to assume no adsorption rate enhancement due to liquid advection.

Nomenclature

Roman

A	Surface interaction parameter	[–]
C	Concentration at the subsurface	[mol.m^{-3}]
C_b	Bulk concentration	[mol.m^{-3}]
d	Diameter of a spherical particle (or bubble)	[m]
D	Coefficient of molecular diffusion	[m^2.s^{-1}]
k	A constant in the Freundlich isotherm	[mol$^{(N-1)/N}$.m$^{(3-2N)/N}$]
K_F	A constant in the Frumkin isotherm	[m^3.mol^{-1}]
K_H	Henry's law constant	[m]
k_L	Fluid film mass transfer coefficient	[m^2.s^{-1}]
K_L	A constant in the Langmuir isotherm	[m^3.mol^{-1}]
K_v	A constant in the Volmer isotherm	[m^3.mol^{-1}]
n	A number taking the value of between 1 and 2	[–]
N	Consistency index in the Freundlich isotherm	[–]
r	Radius of curvature of a spherical interface	[m]
R	Universal gas constant	[J.K^{-1}.mol^{-1}]
Re	Reynolds number defined in Equation 2.20	[–]
Sc	Schmidt number defined in Equation 2.21	[–]
Sh	Sherwood number defined in Equation 2.22	[–]
t	Time	[s]
T	Absolute temperature	[K]
V	Gas–Liquid slip velocity	[m.s^{-1}]

Greek

Γ	Surface excess	[mol.m^{-2}]
Γ_∞	Saturation surface excess	[mol.m^{-2}]
Γ^*	Equilibrium surface excess	[mol.m^{-2}]
μ	Liquid dynamic viscosity	[Pa.s]
Π	Surface pressure	[N.m^{-1}]
ρ_f	Liquid density	[kg.m^{-3}]
σ	Surface tension	[N.m^{-1}]
σ_0	Tension of a surfactant-free surface	[N.m^{-1}]
τ	Dummy variable with units of time	[s]

References

Borwankar, R.P., and Wasan, D.T. 1983. The kinetics of adsorption of surface active agents at gas–liquid surfaces, *Chemical Engineering Science*, 38, 1637–1649.

Brown, A.K., Kaul A., and Varley, J. 1999. Continuous foaming for protein recovery. Part I. Recovery of β-casein, *Biotechnology & Bioengineering*, 62, 278–279.

Burnett, C.R., and Stevenson, P. 2012. Denaturation of proteins upon adsorption to bubble surfaces, Paper 2, *CHEMECA '12*, Wellington, New Zealand, 23–26 September.

Clarkson, J.R., Cui, Z.F., and Darton, R.C. 1999a. Protein denaturation in foam. I. Mechanism study, *Journal of Colloid & Interface Science*, 215, 323–332.

Clarkson, J.R., Cui, Z.F., and Darton, R.C. 1999b. Protein denaturation in foam. II. Surface activity and conformational change, *Journal of Colloid & Interface Science*, 215, 333–338.

Cussler, E.L. 2009. *Diffusion: Mass Transfer in Liquid Systems*, Cambridge University Press, ISBN: 0521871212.

Du, L., Prokop, A., and Tanner, R.D. 2003. Effect of pH on the startup of a continuous foam fractionation process containing ovalbumin, *Separation Science & Technology*, 38, 1093–1109.

Dukhin, S.S., Kretzschmar, G., and Miller, R. 1995. *Dynamics of Adsorption at Liquid Interfaces: Theory, Experiment, Application*, Elsevier, ISBN: 0444881174.

Dukhin, S.S., Miller, R., and Kretzschmar, G. 1983. On the theory of adsorption-kinetics of ionic surfactants at fluid interfaces. 1. The effect of the electric double-layer under quasi-equilibrium conditions on adsorption-kinetics, *Colloid and Polymer Science*, 261, 335–339.

Dukhin, S.S., Miller, R., and Loglio, G. 1998. Physico-chemical hydrodynamics of rising bubble, in *Drops and Bubbles in Interfacial Research*, Elsevier, ISBN: 0444828941.

Eastoe, J., and Dalton, J.S. 2000. Dynamic surface tension and adsorption mechanisms of surfactants at the air-water interface, *Advanced Colloid & Interface Science*, 84, 103–144.

Graham, D.E., and Phillips, M.C. 1979a. Proteins at liquid interfaces. I. Kinetics of adsorption and surface denaturation, *Journal of Colloid & Interface Science*, 70, 403–414.

Graham, D.E., and Phillips, M.C. 1979b. Proteins at liquid interfaces. II. Kinetics of adsorption and surface denaturation, *Journal of Colloid & Interface Science*, 70, 415–426.

Karakashev, S., Nguyen, A., and Miller, J. 2008. Equilibrium adsorption of surfactants at the gas–liquid interface, in *Interfacial Processes and Molecular Aggregation of Surfactants*, Springer, ISBN: 3540698104.

Kralchevsky, P.A, Danov, K.D., Broze, G., and Mehreteab, A. 1999. Thermodynamics of ionic surfactant adsorption with account for the counterion binding: Effect of salts of various valency, *Langmuir*, 15, 2351–2365.

Levich, V. 1962 *Physiochemical Hydrodynamics*, Prentice-Hall, Upper Saddle River, NJ.

Li X., Shaw R., Evans G.M., and Stevenson P. 2010. A simple numerical solution to the Ward-Tordai equation for the adsorption of non-ionic surfactants, *Computers and Chemical Engineering* 34, 146–153.

Liu, Z., Liu, Z., Wang, D., Ding, F., and Yuan, N. 1998. On the denaturation of enzymes in the process of foam fractionation, *Bioseparations*, 7, 167–174.

Makievski, A.V., Fainerman, V.B., Miller, R., Bree, M., Liggieri, L., and Ravera, F. 1997. Determination of equilibrium surface tension values by extrapolation via long time approximations, *Colloids & Surfaces A*, 122, 269–273.

Makievski, A.V., Loglio, G., Krägel, J., Miller, R., Fainerman, V.B., and Neumann, A.W. 1999. Adsorption of protein layers at the water/air interface as studied by axisymmetric drop and bubble shape analysis, *Journal of Physical Chemistry B*, 103, 9557–9561.

Matsumoto, Y., Uda, T., and Takagi, S. 2006. The effect of surfactant on rising bubbles, in *IUTAM Symposium on Computational Approaches to Multiphase Flow*, Springer, ISBN: 1402049774.

Miller, R., Dukhin, S.S., and Kretzschmar, G. 1985. On the theory of adsorption-kinetics of ionic surfactants at fluid interfaces. 2. Numerical-calculations of the influence of a quasi-equilibrium electric double-layer, *Colloid & Polymer Science*, 263, 420–423.

Miller, R., Fainerman, V.B., Wustneck, R., Kragel, J., and Trukhin, D.V. 1998. Characterisation of the initial period of protein adsorption by dynamic surface tension measurements using different drop techniques, *Colloids & Surfaces A*, 131, 225–230.

Miller, R., and Kretzschmar, G. 1980. Numerical solution for a mixed model of diffusion kinetics-controlled adsorption, *Colloid & Polymer Science*, 258, 85–87.

Mysels, K.J. 1982. Diffusion-controlled adsorption kinetics: General solution and some applications, *Journal of Physical Chemistry B*, 86, 4648–4651.

Phan, C.M., Nguyen, A.V., and Evans, G.M. 2005. Dynamic adsorption of sodium dodecylbenzene sulphonate and Dowfroth 250 onto the air–water interface, *Minerals Engineering*, 18, 599.

Sharifullin, V.N., and Luebbert, A. 2001. Adsorption processes and the velocity of a bubble rising in the presence of surfactants, *Theoretical Foundations of Chemical Engineering*, 35, 357–360.

Sillero, A., and Maldonado, A. 2006. Isoelectric point determination of proteins and other macromolecules: Oscillating method, *Computers in Biology & Medicine*, 36, 157–166.

Stejskal, E.O., and Tanner, J.E. 1965. Spin diffusion measurements: Spin echoes in the presence of time-dependent field gradient, *Journal of Chemical Physics*, 42, 288–292.

Tosun, I. 2002. *Modelling in Transport Phenomena: A Conceptual Approach*, Elsevier, ISBN: 0444510525.

Wakao, N., and Kaguei, S. 1982. Heat and mass transfer in packed beds, in *Gordon and Breach Science*, ISBN: 0677058608.

Ward, A.F.H., & Tordai, L. (1946). Time-dependence of boundary tension of solutions I. The role of diffusion in time-effects. *The Journal of Chemical Physics*, 14, 453.

Wilke, C.R., and Chang, P. 1955. Correlation of diffusion coefficients in dilute solutions, *AIChE Journal*, 1, 264–270.

Ybert, C., and di Meglio, J.M. 1998. Study of protein adsorption by dynamic surface tension measurements: Diffusive regime, *Langmuir*, 14, 471–475.

Young, M.E., Carroad, P.A., and Bell, R.L. 2004. Estimation of diffusion coefficients of proteins, *Biotechnology & Bioengineering*, 22, 947–955.

Zhang, Y., McLaughlin, J.B., and Finch, J.A. 2001. Bubble velocity profile and model of surfactant mass transfer to bubble surface, *Chemical Engineering Science*, 56, 6605–6616.

3

Hydrodynamics of Pneumatic Foam

3.1 Traditional Approaches to Foam Fractionation Modeling

Foam fractionation is a physically complex process, and it is for this reason that many analyses of it have tended to overlook the individual physical phenomena that occur but instead look for a more qualitative description. Certainly until 2008 there appears to have been no attempt to explain the process in the context of a detailed description of foam drainage rate other than the analyses by Lemlich and co-workers. However, along with knowledge of how and why some species adsorb to a gas–liquid interface, an understanding of foam hydrodynamics is a pillar upon which innovation in foam fractionation must be based.

Before explaining the hydrodynamics of pneumatic foam it is worth considering how most foam fractionation researchers traditionally, but incorrectly, describe the process. It is a common observation that the foam in a fractionation column becomes monotonically drier from the bottom to the top. Because, all things being equal, a dry foam is desirable to engender high enrichment, the phenomenon is a boon to foam fractionation. The cause of the drying of the foam with height is almost always explained by the fact that the higher an element of the foam goes in the column, the longer it has had to drain. This was described by Stevenson and Li (2012) as the "vertical foam misapprehension," but it has been repeated, as if a mantra, by countless foam fractionation workers. For example, Boonyasuwat et al. (2003) stated that "An increase in foam height leads to a longer foam residence time, which allows more drainage," and Saleh et al. (2006) stated that "Increasing the gas-flow rates resulted in higher volume of wet foam due to the short residence time for the foam to drain the liquid, leading to an increase in recovery but decreased enrichment ratios." In a review of foam separations, Kruglyakov et al. (2008) used observations that the liquid fraction was a function of height to incorrectly infer that foam drainage expressions used by others were incorrect. Stevenson and Li (2012) gave a *reductio ad absurdum* argument to show that the vertical flow misapprehension is the result of an incomplete understanding of foam hydrodynamics.

This chapter will show that the diminution of liquid fraction with height is due to either capillarity (a mechanism of only marginal interest in the context of foam fractionation) or to changes in the bubble size distribution, and the rigorous analysis of hydrodynamic foam will form the basis for a genuine model of the foam fractionation process.

3.2 Foam Drainage

The most fundamental aspect of an understanding of the hydrodynamics of foam is the rate at which liquid drains past the matrix of bubbles through the foam. Foam drainage has been the subject of intensive study over the past four decades, and many of the theories that have been promulgated are so mathematically convoluted as to render them almost impenetrable. Nonetheless, the issue of foam drainage must be tackled. In this section, the traditional models of foam drainage are briefly discussed before a more accessible empirical description is presented that will facilitate a more general description of the hydrodynamics of a pneumatic foam.

Liquid drains through a pneumatic foam under the weight of the liquid contained therein. The drainage is retarded by a shear stress that is imparted by the gas–liquid interfaces. For a foam that is isotropic with respect to global liquid fraction (i.e., the liquid content is constant throughout the foam), these are the only two forces that act upon the liquid and therefore they solely govern the drainage rate. However, in a foam where the global liquid fraction changes in one direction or another, there exist capillary forces that are imparted upon a liquid that are caused by a finite surface tension and global changes in the radius of curvature of the channels that contain the liquid. Such capillary forces are demonstrated in the photographs of Figure 3.1. Liquid is added via a needle to the bottom of a very dry foam; the initial condition is shown in the left-hand frame. It is the capillary forces that draw the liquid up the column (i.e., in the opposite direction of gravity) to create a wetter foam.

However, before commencing a description of foam drainage, it is appropriate to briefly consider the structure, or morphology, of foam. When foam is very wet (i.e., it has a very high liquid fraction), the bubbles manifest as spheres and the liquid fraction can approach that of the Bernal close-packing limit (i.e., approximately 36%; Bernal and Mason, 1960). However, as the liquid fraction diminishes from this value, the structure demonstrates the characteristics of liquid-containing channels that meet at junctions. These channels have a sectional shape of the convex tricuspid that are known as Plateau borders, of which four meet at nodes (demonstrated in Figure 3.2). A photograph of a pneumatic foam taken through a transparent planar column wall is given in Figure 3.3.

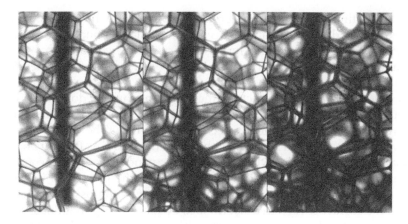

FIGURE 3.1
Demonstration of the capillary effect in a dry foam drawing liquid against the direction of gravity.

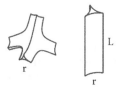

FIGURE 3.2
A node of foam structure (left) where four Plateau borders meet. A single Plateau border (right) of length L and the cross-sectional shape of a convex tricuspid of radius of curvature r. (Courtesy of Stephan Koehler.)

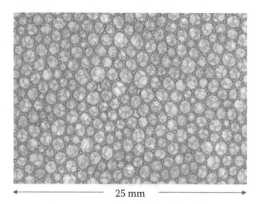

25 mm

FIGURE 3.3
A photograph of pneumatic foam of liquid fraction of approximately 4% taken through a planar transparent column wall.

In the wet limit, the Plateau borders are indistinct from the nodes; but as the foam becomes drier, the channels become more slender and the distinction becomes clear, and it is this that forms the basis of the two traditional schools of foam drainage modeling: channel-dominated and node-dominated foam drainage.

As described in the introduction, the theory of channel-dominated foam drainage was first promulgated by Leonard and Lemlich (1965) and subsequently, although independently, recreated by others. The theory assumes that viscous losses occur entirely within the channels (i.e., Plateau borders) and not within the nodes where the liquid expands, bends, and contracts. Crucially, Leonard and Lemlich recognized that the gas–liquid interfaces were not completely rigid, and suggested that the viscous losses were a function of the surface shear viscosity. (In fact, by recognizing that Boussinesq's (1913) theory for the force on a bubble rising at low Reynolds number is a function of surface dilatational viscosity, it is likely that the viscous losses are governed by the surface shear viscosity in the dry limit, but that the surface dilatational viscosity is a contributing factor for wetter foams. For an excellent description of surface rheology, see the book of Edwards et al. (1991).)

A counter-proposal, node-dominated foam drainage, was made by Koehler et al. (1999), who assumed that the viscous losses occurred wholly at the nodes and that the gas–liquid interface within the channels could not support a shear stress, and therefore viscous losses. Koehler et al. (2000) subsequently presented a theory that allowed for losses in both the nodes and in the channels.

However, Stevenson (2006) attempted to circumvent the mathematical complexity of previous drainage theories and performed a dimensional analysis to relate the superficial drainage velocity in a Lagrangian frame of reference moving with the bubbles, j_d, to a representative bubble radius, r_b; the liquid density and dynamic viscosity, ρ_f and μ, respectively; and the acceleration due to gravity, g. No attempt was made to relate the superficial drainage velocity to the surface shear viscosity because the measurement of the latter quantity is exceedingly unreliable (Stevenson, 2005) and therefore the dependency of drainage upon surface rheology is described via adjustable constants. Making the assumption that losses due to drainage were entirely viscous (and not inertial), it was shown that the drainage rate, nondimensionalized as a "Stokes number," in an isotropic foam is a function of liquid fraction only, i.e.,

$$Sk = \frac{\mu j_d}{\rho_f g r_b^2} = f(\varepsilon) \tag{3.1}$$

Further, it was suggested that a power-law relationship can well-approximate drainage data, that is,

$$Sk = m\varepsilon^n \tag{3.2}$$

TABLE 3.1

Drainage Equations in the Form of Equation 3.2 Fitted to Previous Data Sets

Source	Solution	Bubble Diameter (mm)	Drainage Equation
Weaire et al. (1993)	"Detergent"	0.77	$SK = 0.046\varepsilon^{2.13}$
Neethling et al. (2002)	SDS (conc. unspecified)	1.6 and 2.5	$SK = 0.012\varepsilon^{1.74}$
	TTAB (conc. unspecified)	1.6 and 2.5	$SK = 0.013\varepsilon^{1.78}$
Stevenson et al. (2007a)	2.92 g.L^{-1} SDS in 0% and 40% glycerol	0.58 and 0.80	$SK = 0.016\varepsilon^{2.00}$
Koehler et al. (1999)	0.25 wt% Dawn Soap™	5.56 est.	$SK = 0.0011\varepsilon^{1.56}$
		2.36 est.	$SK = 0.0022\varepsilon^{1.56}$
		1.42 est.	$SK = 0.0027\varepsilon^{1.55}$
Durand et al. (1999)	0.25 wt% Dawn Soap™	1.00 approx.	$SK = 0.052\varepsilon^{1.64}$
	3.46 g.L^{-1} SDS, 1.73 mg.L^{-1} Dodecanol	1.00 approx.	$SK = 0.076\varepsilon^{2.17}$
	3.46 g.L^{-1} SDS, 0.86 mg.L^{-1} Dodecanol	1.00 approx.	$SK = 0.021\varepsilon^{1.64}$
Saint-Jalmes and Langevin (2002)	SDS (conc. unspecified conc) with Dodecanol (at conc. 250 times less than SDS)	0.80	$SK = 0.060\varepsilon^{2.12}$
		0.18	$SK = 0.011\varepsilon^{1.45}$

where m and n are adjustable constants that are implicit functions of surface parameters, and can be measured by the method of forced drainage (Weaire et al., 1993) or magnetic resonance imaging (Stevenson et al., 2007a), the latter who gave the parameters $m = 0.046$ and $n = 2.00$ for a foam stabilized by 2.92 g.L^{-1} sodium dodecyl sulfate. Stevenson (2008) surveyed the available drainage data in the literature to which the parameters m and n were fitted; the collated data are given in Table 3.1.

For a foam with a spatially variant liquid fraction with height in the column, Stevenson (2006) suggested the following modification to Equation 3.2 to allow for capillarity:

$$Sk = m\varepsilon^n (1 + \Pi) \tag{3.3}$$

where

$$\Pi = \frac{\sigma}{2.78\, r_b\, \rho_f\, g\, \varepsilon^{1.46}} \frac{\partial \varepsilon}{\partial x} \tag{3.4}$$

and x is the vertical distance measured positive upward.

This approach to describing foam drainage is overtly empirical in that it requires the use of two adjustable constants that must be found experimentally. However, the combined drainage theory of Koehler also requires two adjustable constants, although they are not called such. The simplicity of

Equation 3.2 makes drainage theory tractable and amenable to further use to describe the hydrodynamics of pneumatic foam in the next section.

Finally, it should be noted that the dimensionless analysis that led to Equation 3.1 assumes that pressure losses due to liquid flow through the foam are entirely viscous and that inertial losses are insignificant. An extension to account for inertial losses has been proposed by Stevenson and Li (2010), but such losses are only significant in very wet foam that could not be practicably applied in foam fractionation operations.

3.3 The Hydrodynamic Theory of Pneumatic Foam

In a foam fractionation column, the foam rises upward until it reaches the top and is removed either via an inverted U-piece or a froth launder (see Section 7.4). As the foam rises through the column, the bubble size distribution may change, either due to inter-bubble gas diffusion or film rupture (see Sections 4.2 and 4.3, respectively). If a froth launder is employed to remove the foam, then there may be film rupture of the free surface of the form. In addition, the bubble size distribution may change because of a reduction in the static pressure upward through the foam column, although this is always insignificant compared to atmospheric pressure and will be overlooked.

Recall that, in foam fractionation, the flux of surface is desirable but the flux of interstitial liquid is not, and therefore one must seek a descriptor of the flux of both quantities; this can be conveniently done using the "Hydrodynamic Theory of Pneumatic Foam" developed by Stevenson (2007b).

The liquid flux (or superficial velocity), j_f, up the column can be written as

$$j_f = \frac{\varepsilon j_g}{1-\varepsilon} - j_d \qquad (3.5)$$

where j_g is the gas superficial velocity. By substituting Equation 3.1 for the drainage term into Equation (3.5),

$$j_f = \frac{\varepsilon j_g}{1-\varepsilon} - \frac{\rho_f g r_b^2}{\mu} m \varepsilon^n \qquad (3.6)$$

which gives a locus of j_f versus ε, which for typical values of the relevant parameters are plotted in Figure 3.4.

By stability analysis, it can be shown that the equilibrium condition of the pneumatic foam is given by the peak of the curve shown in Figure 3.4. It is seen that the liquid fraction, ε, i.e., the equilibrium liquid fraction, is calculated by numerical solution of the expression

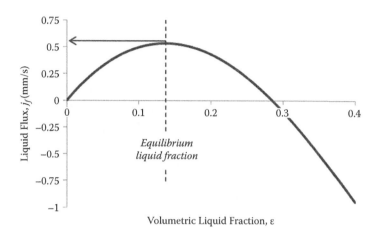

FIGURE 3.4

The locus of liquid superficial velocity $\mu = 1.0$ mPa.s, $\rho_f = 1,000$ kg.m^{-3}, $j_g = 8$ mm.s^{-1}, $r_b = 0.5$ mm, $m = 0.016$, and $n = 2$. The equilibrium liquid fraction is seen to be 0.137 (i.e., 13.7%), which gives $j_f = 0.533$ mm.s^{-1}.

$$\frac{\mu j_g}{mn\rho_f g r_b^2} = \varepsilon^{n-1}(1-\varepsilon)^2 \tag{3.7}$$

and the liquid flux is calculated by substituting the value of ε into Equation 3.6.

It is seen from Figure 3.5 that the equilibrium liquid fraction increases monotonically with increasing gas rate, and this observation is crucial for the design of a foam fractionation process. However, it can be stated that, at this stage, the hydrodynamic performance of a column of pneumatic foam, and therefore the performance of a continuous foam fractionation operation, is most easily controlled by varying the gas rate or by manipulating the bubble size. To demonstrate this, consider Figures 3.5 and 3.6, which show the dependency of the liquid fraction and liquid superficial velocity on gas superficial velocity and bubble size, respectively, for the conditions $\mu = 1.0$ mPa.s, $\rho_f = 1,000$ kg.m^{-3}, and $m = 0.016$, $n = 2$. In Figure 3.5, the bubble radius, r_b, is set at 0.5 mm and the gas superficial velocity is varied. According to the theory expounded herein, both the liquid fraction and the liquid flux increase with increasing gas rate. In Figure 3.6, the gas superficial velocity is set at 1.0 mm.s^{-1} and the hydrodynamic condition of the pneumatic foam calculated as a function of bubble radius. Both liquid fraction and liquid flux increase with decreasing bubble size, with dramatic variation observed below a value of r_b of around 0.2 mm (200 µm). It will be seen subsequently that the strongest control parameter for the performance of foam fractionation is the bubble size.

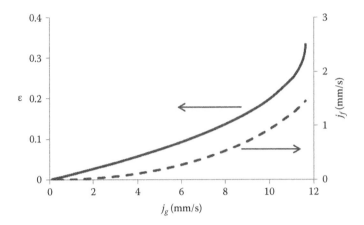

FIGURE 3.5
The variation of liquid fraction and liquid flux according to the theory of Stevenson (2007b) with gas superficial velocity j_g under the conditions of $\mu = 1.0$ mPa.s, $\rho_f = 1,000$ kg.m^{-3}, $m = 0.016$, $n = 2$, and $r_b = 0.50$ mm.

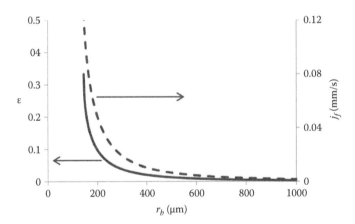

FIGURE 3.6
The variation of liquid fraction and liquid flux according to the theory of Stevenson (2007b) with bubble radius r_b under the conditions of $\mu = 1.0$ mPa.s, $\rho_f = 1,000$ kg.m^{-3}, and $m = 0.016$, $n = 2$, and $j_g = 1.0$ mm.s^{-1}.

It should be noted that, when foam fractionators are run as enrichers, it is usual to run with a very dry foam to promote enrichment, as will be discussed in Chapters 6 and 7. Certainly, a liquid fraction of 13.7%, as is developed in Figure 3.4, would be considered very wet for an enricher, although such liquid fractions may be seen in strippers.

3.4 Nonoverflowing Pneumatic Foam

The simplest, but not necessarily most practical, method of performing an enrichment by foam fractionation is to bubble gas through a liquid such that the foam rises up a column until it reaches an equilibrium height. (As an aside, it is noted that the measurement of an equilibrium foam height by this method is the basis of the Bikerman (1938) foam stability test.) An equilibrium height is attained because the bubbles at the surface rupture (some might have also burst within the bulk of the foam, known as "internal coalescence"), and there is zero net liquid flux up the column (i.e., $j_f = 0$), which is identical to the state of "total reflux" that will be described in Chapter 6. When the foam is formed on the top of the column, liquid can be introduced at the bottom of the column to expel the foam. The enrichment engendered through this method was measured in the device described in Figure 9.3, but it is appropriate at this juncture to describe this particular system in the context of the hydrodynamic theory described above. Because the liquid flux is zero, Equation 3.7 can be equated to zero, that is,

$$j_f = \frac{\varepsilon j_g}{1-\varepsilon} - \frac{\rho_f g r_b^2}{\mu} m \varepsilon^n = 0 \tag{3.8}$$

So, if capillary effects are overlooked, the equilibrium liquid fraction can be calculated by numerical solution of Equation 3.8. In the special case that $n = 2$, Equation 3.8 can be solved analytically to give

$$\varepsilon = \frac{1}{2} - \sqrt{\frac{1}{4} - \frac{\mu j_g}{m \rho_f g r_b^2}} \tag{3.9}$$

3.5 Toward a Single-Parameter Description of Bubble Size from a Size Distribution

In the above analysis, a single-parameter description of the bubble radius, r_b, has been adopted. However, a foam of uniform bubble diameter is almost never produced during a foam fractionation operation. Certainly, when foam is produced by sparging through a glass frit or perforated plate (see Section 7.2 for a discussion about bubble formation), a polydispersive bubble size distribution is exhibited. For example, Figure 3.7 shows the cumulative bubble diameter distribution of a pneumatic foam stabilized by the ionic surfactant CTAB formed by sparging gas through a perforated tube.

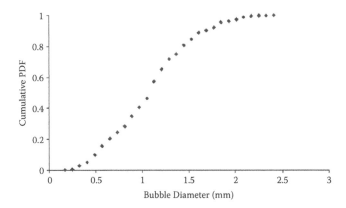

FIGURE 3.7
Cumulative bubble diameter distribution of a pneumatic foam stabilized by CTAB. (Data courtesy of Matthew Patterson.)

The distribution was measured by photography through a planar transparent column wall. The poly-dispersivity is self-evident.

Thus, the question arises as to what is the appropriate size of bubble to describe the distribution in the drainage law of Equation 3.2. This is a particularly vexed question because it is very difficult to obtain a reliable bubble size distribution, and, even if such a distribution is available, it is not immediately apparent what single-parameter description is appropriate.

The difficulty in measuring the bubble size distribution is such that the vast majority of foam fractionation studies do not endeavor to measure it, and instead report an approximate average bubble size. For instance, in the forced drainage experiments of Durand et al. (1999), they reported a mean bubble diameter of "approximately 1 mm," which suggests that not a great deal of importance was placed on measuring this parameter. Given that the hydrodynamic condition of a pneumatic foam is so strongly dependent on bubble size (see Figure 3.6), the performance of foam fractionation is critically determined by the bubble size distribution. Lemlich certainly appreciated the importance of bubble size distribution and he studied the most common experimental method of measuring the distribution (Cheng and Lemlich, 1983), which is to measure the distribution photographically through a transparent planar wall in the column. The point was made that the distribution taken through the column walls is not necessarily representative of the size distribution within the bulk of the foam because

1. Statistical planar sampling bias discriminates against the sampling of smaller bubbles, but
2. The small bubbles tend to wedge the larger bubbles away from the wall, and
3. The column wall causes bubble deformation.

The first of these problems can be rectified mathematically: Let $f(r_b)$ be the bubble radius distribution taken by observation through a plane, and $g(r_b)$ the distribution within a volume in the bulk of the foam. It is the latter distribution that is of practical utility for the purposes of describing the hydrodynamic condition in foam fractionation. de Vries (1972) showed that the two distributions are related by the expression.

$$g(r_b) = \frac{f(r_b) \int_0^\infty r_b f(r_b) \, dr_b}{r_b} \tag{3.10}$$

There are no methods to correct the second and third points of Cheng and Lemlich (1983). The best that can currently be done is to suppose that points 1 and 2 self-cancel and that point 3 is insignificant. Given that bubble size distribution is so important to the performance of foam fractionation, this situation is less than ideal.

Clearly, a method of directly measuring the bubble size distribution with the bulk of a pneumatic foam would be desirable. The closest approach to this goal is due to Stevenson et al. (2010), who used a Pulsed-Field Gradient Nuclear Magnetic Resonance (PFG-NMR) technique to measure the size distribution in a non-overflowing pneumatic foam; an example for a foam stabilized by the ionic surfactant SDS created by sparging gas through a glass frit is given in Figure 3.8. However, this technique has its limitations in that it cannot presently obtain distributions from a practical overflowing foam such as those typical in foam fractionation; they require sparging with a

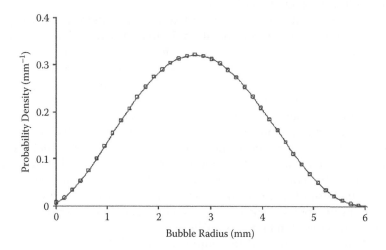

FIGURE 3.8
An example of a bubble size distribution from the bulk of a foam stabilized by SDS using PFG-NMR.

magnetically susceptible gas (such as butane or pentane) and the method requires the use of a specialist high-field magnetic resonance spectrometer.

The lack of practical method for measuring bubble size distribution within the bulk of a pneumatic foam is indeed problematical. However, should such a method become available, it is appropriate to consider what single-parameter descriptor of a poly-dispersive size distribution, $\bar{r}_b$, is appropriate for use in the drainage law. There is no experimentally verified theory for such a single parameter, and the best that can be done is to offer the unverified hypothesis, derived in the appendix to this chapter, that

$$
\bar{r}_b = \left[\frac{\displaystyle\int_0^\infty r_b^5 g(r_b)\,\mathrm{d}r_b}{\displaystyle\int_0^\infty r_b g(r_b)\,\mathrm{d}r_b} \right]^{1/4}
\tag{3.11}
$$

3.6 Hydrodynamic Modulation Due to "Washwater" Addition

In order that gangue material can be effectively rejected from the concentrate stream in column flotation, washwater is routinely sprayed onto the top of the column. If the rate of washwater addition is greater than the flux of liquid from the top of the column that would be apparent if there was no washwater, there is a net flux of liquid downward through the foam, and the column is said to be in positive bias, which is essential to achieve excellent gangue rejection (Stevenson et al., 2007).

There is, yet again, an operation in foam fractionation that corresponds to washwater addition in froth flotation: In foam fractionation with *external reflux*, described fully in Section 6.5, used to increase the enrichment factor, some of the collected foamate is added as washwater. This has a similarity to the return of condensed overhead product back to the top of a distillation column as a reflux stream.

It is of present interest to determine the hydrodynamic condition due to the addition of reflux, and this can be done by modifying the construction given in Figure 3.4. It has been shown (Stevenson et al., 2007) that the addition of washwater into a mature pneumatic foam (i.e., one that has temporally attained the equilibrium condition) will move down the column and enhance the liquid fraction of the foam, but the condition of the foam continues to reside upon the locus given by Equation 3.7. For example, take the foam first considered in Figure 3.4. When the foam is mature, the liquid fraction is 0.137 (i.e., 13.7%) and the liquid flux is 0.533 mm.s^{-1}. However, if a flux of 0.3 mm.s^{-1} is returned to the column, all reflux liquid travels down the column, resulting in a net liquid flux up the column of 0.233 mm.s^{-1}. This new liquid flux is drawn as the vertical dashed line in Figure 3.9.

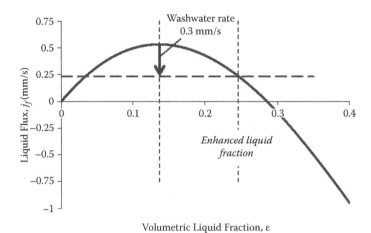

FIGURE 3.9
Modulation of the hydrodynamic condition of a pneumatic foam by washwater addition/ external reflux.

The equilibrium liquid fractionation enhances to 0.246 (i.e., 24.6%) and the liquid flux removed from the system is 0.233 mm.s^{-1}.

If all the foamate liquid were returned to the column (i.e., at a flux of 0.533 mm.s^{-1}), then the net flux would be zero and the column would be said to be operating at *total reflux*, which is equivalent to incipient positive bias in flotation.

3.7 Hydrodynamic Modulation Due to Bubble Disproportionation

As discussed in Section 3.1, the *vertical foam misapprehension* has it that the liquid fraction of a vertical column of pneumatic foam diminishes with height because foam at a higher level has had longer to drain. This assertion has no basis in any mass balance that can be drawn on the system. In fact, the liquid fraction decreases because of (1) the relaxation of capillary forces at the bottom of the foam layer, which for practical foam fractionation design purposes can be discounted; and (2) the change in bubble size distribution with height (i.e., "disproportionation") caused by Ostwald ripening (see Section 4.2) and liquid film rupture (Sections 4.3 and 4.4). Due to the effects of Ostwald ripening, the characteristic bubble radius r_b increases with height.

For the purposes of this section the physical mechanisms of disproportionation will not be considered; rather, the effect of monotonically increasing

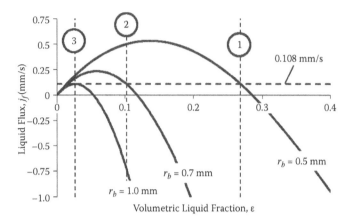

FIGURE 3.10
The construction required to calculate the change in the hydrodynamic condition of a pneumatic foam due to bubble disproportionation with height.

the value of r_b with height on the hydrodynamic condition of a steady-state pneumatic foam will be considered. Throughout the analysis, the effects of capillarity will be discounted.

Consider a pneumatic foam with μ = 1.0 mPa.s, ρ_f = 1,000 kg.m^{-3}, and m = 0.016, n = 2, and j_g = 8 mm.s^{-1}. The characteristic bubble radius r_b at the bottom of the foam (State "1") is 0.5 mm but disproportionation increases the radius to 0.7 mm at an intermediate location (State "2"); and at the top of the column, r_b is 1.0 mm.s^{-1}. The net flux of liquid up the column is given by the equilibrium condition at the top of the column: In Figure 3.10, the locus of j_f-ε relevant to r_b = 1.0 mm.s^{-1} is drawn, and the equilibrium condition is given by the peak of the curve. The equilibrium liquid fraction at the top of the column is seen to be 0.027 (i.e., 2.7%), and the liquid flux leaving the top of the column is 0.108 mm.s^{-1}. However, at steady state, this is the liquid flux that manifests at all locations in the column. Thus, the hydrodynamic state at locations below the top of the column is given by right-hand side intersection of the line j_f = 0.108 mm.s^{-1} (i.e., the foam becomes wetter) with the locus that is pertinent to characteristic bubble radius at that position. Thus, it is seen that at the intermediate position (State "2"), the liquid fraction is 0.102 (i.e., 10.2%), and at the bottom of the column (State "3") the liquid fraction is 0.268 (i.e., 26.8%). In fact, it is seen that, when the top of the column opens for the foam to be collected in a "launder vessel" (see Section 7.4), there can be significant bubble rupture on the free surface, and it is the characteristic bubble radius at the free surface that is the relevant one with which to construct the top locus. Such considerations are not required when the foam is collected using an inverted "U" arrangement (see Section 7.4).

It is seen that, as bubbles increase their characteristic size through disproportionation, liquid that was previously traveling up the column is liberated to provide washwater to the foam beneath, thereby enhancing the liquid fraction in a manner similar to that seen in Section 3.6 when washwater is added to the top of the column. This forms the basis of the *internal reflux* mechanism that is discussed in Section 6.7.

3.8 Start-Up Transients

The above mechanistic descriptions of a pneumatic foam have assumed that the column is at steady state and therefore does not vary in time. The attainment of steady state is of particular importance when operating a column under external reflux because liquid delivered to the top of the column will only transport downward upon attaining steady state (i.e., hydrodynamical maturity). Shaw et al. (2011) showed that a foam created with a very high gas flux of 30 mm.s^{-1} could take up to 3 minutes to equilibrate, but foam fractionation columns that operate at typically much lower gas rates can take considerably longer.

3.9 Foam Flowing through Columns of Varying Cross-Sectional Area

A device for the process intensification of foam fractionation that forces the rising foam through successive contraction and expansion stages is described in Section 9.6, so it is pertinent to consider how changes in the cross-sectional area of the column modify the hydrodynamic state of the foam. The theory for such modification has been explained in full by Li et al. (2011). Again, the changes in the hydrodynamic state will be explained by appealing to graphical construction. It will be seen that squeezing pneumatic foam through a contraction (somewhat counter-intuitively) causes the liquid fraction to diminish but the volumetric liquid flowrate to remain constant, whereas when a foam is vertically expanded, the liquid fraction again diminishes but liquid is rejected from the foam, causing a reduction in volumetric flowrate. To put this another way, *any* change in cross-sectional area causes a reduction in liquid fraction, and only expansions create a lower liquid flowrate.

First we discuss contractions of a cross-sectional area. Consider a foam flowing through a sudden contraction in a vertical tube where the flow area is halved. The superficial gas and liquid velocities above the contraction will be twice as high as those below the contraction, by continuity. Assuming no changes in bubble size distribution, liquid superficial velocity versus liquid fraction curves for the foam described above are given in Figure 3.11.

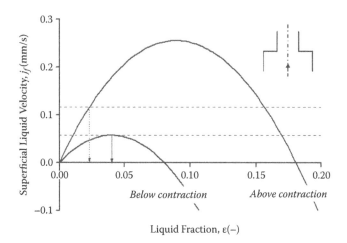

FIGURE 3.11
Liquid flux versus liquid fraction curves for the foam below and above the contraction, assuming a contraction ratio of 0.5, $\mu = 1$ mPa.s, $\rho_f = 1,000$ kg.m^{-3}, and $j_g = 2.9$ mm.s^{-1} in the main column. The bubble radius $r_b = 0.5$ mm is constant up the column.

The lower horizontal dashed line indicates the liquid superficial velocity before the contraction (0.058 mm.s^{-1}), corresponding to an equilibrium liquid fraction of approximately 0.040 (solid arrow). The upper horizontal dashed line represents the liquid flux above the contraction, which is twice as high as that in the main column. This line has two intersection points with the liquid flux curve for the foam above the contraction, one at a liquid fraction of 0.023 and the other at liquid fraction of 0.16 (i.e., 16%), which is higher than the equilibrium liquid fraction. For the foam to reach the state represented by the right-hand side intersection point (i.e., $\varepsilon = 0.16$), an external reflux of 0.13 mm.s^{-1} must be added. Because external reflux is not employed, only the left-hand side intersection point, $\varepsilon = 0.023$, is a plausible solution. It can be seen that although the superficial gas velocity is doubled above the contraction, the liquid fraction decreases.

This underlying cause of the change of liquid fraction upon contraction can be thought of in a different sense. The liquid fraction ε can be expressed as

$$\varepsilon = \frac{j_f + j_d}{j_f + j_g + j_d} \qquad (3.12)$$

Upon halving the flow cross-sectional area due to the sudden contraction, both j_f and j_g are doubled, whereas j_d stays constant, and this changes the liquid fraction.

Because the reduction in liquid fraction due to the contraction is somewhat counter-intuitive, it is worth providing some experimental evidence

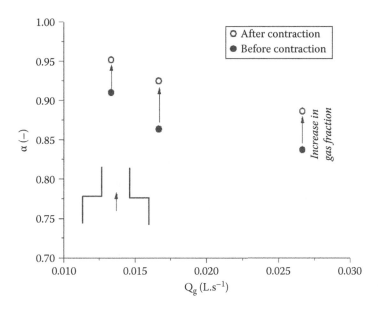

FIGURE 3.12
Experimental results showing that the liquid fraction decreases across the contraction at three different volumetric gas flow rates (0.013, 0.017, and 0.027 L.s^{-1}). The liquid phase was 2.92 g.L^{-1} SDS solution, and the gas phase was air. The contraction ratio is 0.31.

herein, as taken from Li et al. (2011). Figure 3.12 shows the increase in the gas fraction a (which is complementary to the liquid fraction ε) due to a contraction in which the cross-sectional area of the upper tube is 31% of that of the lower tube (the contraction ratio is said to be 31%) for three different volumetric gas flowrates, Q_g. It is seen that, in each case, the gas fraction increases (i.e., the liquid fraction decreases). However, the volumetric liquid flowrate is constant, as it must be through considerations of continuity.

Now consider the case of a sudden expansion. Assuming that the flow area is doubled by the expansion, the liquid flux versus liquid fraction curves can be plotted in a similar manner as in the contraction case above. Note that here the foam-liquid pool interface is in the narrower column below the expansion. Recognize that the volumetric gas and liquid flowrates must be the same below and above the expansion; the liquid flux and liquid fraction in both sections can be determined graphically as shown in Figure 3.13. It can be seen that, above the expansion, both the superficial liquid velocity and the liquid fraction decrease. This provides reflux to the foam in the narrower column, resulting in a higher liquid fraction below (therefore the right-hand side intersection point is the solution). The net liquid superficial velocity in the narrower section is reduced due to the effective reflux.

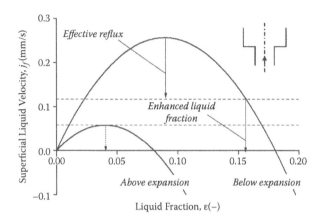

FIGURE 3.13
Liquid flux versus liquid fraction curves for the foam below and above the expansion.

3.10 Beyond the Stable Limit of Foam

High enrichment is typically desirable during foam fractionation, and this generally occurs under conditions of low liquid flux, low gas flux, and low volumetric liquid fraction. However, for completeness, it is worth briefly discussing the behavior of a pneumatic foam at high gas rates. As described in Section 3.3, the equilibrium state of a pneumatic foam in a vertical column is given by the peak of the curve in Figure 3.4, and the liquid fraction increases with increasing gas rate. In such a foam, the foam layer itself is present directly above the bubbly liquid layer. However, as the gas rate increases still further, the distinction between the two layers is eventually lost. Wang et al. (2013) suggested that this condition, known as "flooding," occurs when the locus of j_f-ε no longer exhibits a maximum. This incipiently occurs at a critical gas rate j_g^*:

$$j_g^* = \frac{\rho_f g r_b^2}{\mu} mn \left(\frac{2}{n+1}\right)^2 \left(\frac{n-1}{n+1}\right)^{n-1} \tag{3.13}$$

which corresponds to a maximum liquid fraction of

$$\varepsilon^* = \frac{n-1}{n+1} \tag{3.14}$$

and a maximum liquid flux of

$$j_f^* = \frac{\rho_f g r_b^2}{\mu} m \left(\frac{n-1}{n+1}\right)^{n+1} \tag{3.15}$$

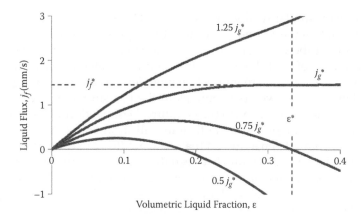

FIGURE 3.14

Equation 3.6 plotted for four different gas rates below, above, and at the critical gas rate. $\mu = 1.0$ mPa.s, $\rho_f = 1{,}000$ kg.m^{-3}, $m = 0.016$, $n = 2$, and $r_b = 0.5$ mm. The dotted lines indicate the liquid fraction and liquid flux at the critical condition. j_g^* in this case is 11.63 mm.s^{-1}.

Figure 3.14 shows loci for four different gas rates. At critical gas rate j_g^*, there is incipiently no maximum exhibited. At gas rates above the flooding condition, gross bubbles can manifest through the column as the "slug flow" regime occurs.

3.11 Wall Shear Stress, Rheology, and Horizontally Flowing Pneumatic Foam

In the above analysis, the contribution of wall shear stress to the pressure gradient in pneumatic foam has been entirely overlooked. In a vertical pneumatic foam, the measured pressure gradient is the sum of two physical factors:

1. The weight of the fluid in the column, which gives rise to a pressure gradient of

$$\rho_f g \varepsilon + \rho_g g (1 - \varepsilon) \approx \rho_f g \varepsilon \tag{3.16}$$

2. The wall shear stress, τ_W, which would give rise to a pressure gradient of

$$\frac{dp}{dx} = \frac{-4\tau_W}{D} \tag{3.17}$$

in a column of circular section of diameter D. However, it has been shown by Li et al. (2011) that the second contributor to the pressure gradient is insignificant for all but the very driest of foams. As can be seen from Equation 3.17, the contribution of wall shear stress to column diameter decreases inversely, so that in industrial scale foam fractionation devices, it is safe to neglect this factor. However, the two contributions to the pressure gradient have been the source of much confusion in the literature; Deshpande and Barigou (2001) falsely attributed the pressure gradient to wall shear stress in its entirety.

In addition, it has been assumed that the foam flows in "plug flow," that is, there is no relative rate of strain between the bubbles and they move upward through the column at a uniform velocity. Foam exhibits a yield stress and if this yield stress is not exceeded, then the bubbles will be unable to slip past one another. Such plug flow behavior has been observed in all of the authors' experiments on foam fractionation.

When observing a foam in plug flow rising through a transparent tube, it is apparent that the bubbles adjacent to the column wall exhibit the same velocity as those toward the center of the column. This situation arises because of "wall-slip," which is quantitatively defined as the velocity of the fluid (in this case, foam) immediately adjacent to the pipe wall, relative to the wall, U_s. In a foam, the cause of the slip velocity is a rarefied liquid layer adjacent to the pipe wall that experiences a high rate of strain, which gives the impression that foam is slipping along the surface, as indicated in Figure 3.15. There are two commonly adopted descriptors of slip velocity:

1. Mooney (1931) postulated that

$$U_s = f(\tau_w) \tag{3.18}$$

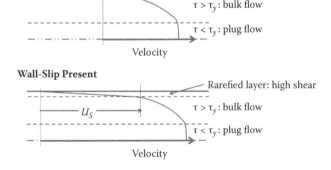

No Slip

$\tau > \tau_y$: bulk flow

$\tau < \tau_y$: plug flow

Velocity

Wall-Slip Present

Rarefied layer: high shear

U_S

$\tau > \tau_y$: bulk flow

$\tau < \tau_y$: plug flow

Velocity

FIGURE 3.15

A schematic of the velocity profile of foam flowing through a vertical tube if there is no slip, and if there is a slip velocity of U_S. (Adapted from Martin, P.J., and Wilson, D.I. 2005. *Chemical Engineering Science*, 60, 493–502.)

and

2. Jastrzebski (1967) postulated that

$$U_S = f\left(\frac{\tau_W}{D}\right) \qquad (3.19)$$

Martin and Wilson (2005) pointed out that the former made more physical sense and determined that the latter gave rise to problematical data interpretations, and thus recommended the use of Mooney's postulate for several classes of colloidal fluids, including foams.

It is worth briefly discussing how one might estimate the wall shear stress should it indeed be considered significant, along with the velocity profiles in a foam should the yield stress be exceeded. To do so, it is first necessary to consider the constitutive relationship between the shear stress, τ, and the rate of plane strain, $\dot{\gamma}$, for a gas–liquid foam. It is generally accepted (Weaire, 2008) that foam exhibits Herschel–Buckley-type rheology, in that it supports a yield stress, τ_y, but a power law relationship beyond this point, i.e.,

$$\tau = \tau_y + k\dot{\gamma}^n \qquad (3.20)$$

It is a common bathing experience that foams with a very low liquid fraction, such as the bubbles formed by adding "bubble-bath" to one's bath water, have a relatively high yield stress (and thus do not shear across one another) but the lather formed from shampooing one's hair has a far lower yield stress and behaves more like a fluid. In Section 7.4, issues surrounding the handling of foamate froth will be discussed, but at this juncture, it is worth mentioning that wet foams are much easier to handle because they flow, whereas dry foams do not.

A quantitative constitutive relationship is difficult to estimate because the three parameters in the Herschel–Buckley expression are strongly dependent on many variables, not the least of which are the bubble size and the liquid fraction. The most practical method of determining the constitutive equation for a foam exhibiting no wall-slip is to measure the flowrate of foam versus pressure gradient in a tube and then apply the Rabinowitsch relationship (adapted to the present nomenclature and assuming no slip between the liquid and gas phases) to link wall shear stress with the strain rate at the wall:

$$\dot{\gamma}_W(\tau_W) = 3\left(j_g + j_f\right) + \frac{dp}{dx} \frac{d\left(j_g + j_f\right)}{d\left(\frac{dp}{dx}\right)} \qquad (3.21)$$

FIGURE 3.16
A photograph of slugs of foam in a horizontal pipe, due to Briceño and Joseph (2003). (From Briceño, M.I., and Joseph, D.D. 2003. *International Journal of Multiphase Flow*, 29, 1817–1831. With permission.)

In fact, wall-slip is almost always present, so the Rabinowitsch equation is modified according to Yilmazer and Kalyon (1991):

$$\dot{\gamma}_W(\tau_W) = \left(j_g + j_f - U_S\right)\left(3 + \frac{d\left[\ln\left(j_g + j_f - V_S\right)\right]}{d\left[\ln \tau_W\right]}\right) \qquad (3.22)$$

Equation 3.21 requires knowledge of the wall-slip velocity, which can be either directly measured experimentally in a foam, or can be deduced as a function of wall shear stress by performing experiments on pipes with multiple diameters and applying the following relationship:

$$U_S(\tau_W) = \frac{1}{2}\frac{\partial\left(j_g + j_f\right)}{\partial\left(\dfrac{1}{D}\right)}\Bigg|_{\tau_W} \qquad (3.23)$$

Finally, a word of caution is required: Briceño and Joseph (2003) defined seven possible flow regimes for the horizontal flow of foam in tubes. Figure 3.16 shows the flow of foam in slug flow. The above analysis is only valid for homogenous flow of foam through a pipe that occurs at relatively high liquid and gas rates.

Nomenclature

Roman

B Geometrical constant employed in Equation (3A.1)   [–]

D Pipe diameter   [m]

g Acceleration due to gravity   [m.s^{-2}]

j_d Superficial liquid drainage velocity   [m.s^{-1}]

j_f Superficial liquid velocity up the column   [m.s^{-1}]

j_f^* Liquid superficial velocity at the flooding condition   [m.s^{-1}]

j_g Superficial gas velocity   [m.s^{-1}]

j_g^* Superficial gas velocity at the flooding condition   [m.s^{-1}]

k Pre-factor in Equation (3.20)   [kg.m^{-1}.s^{n-2}]

m Dimensionless number used in Equation 3.2   [–]

n Dimensionless index used in Equation 3.2, or index in Equation 3.20   [–]

p Pressure   [N.m^{-2}]

Q_g Volumetric gas flow rate   [m^3.s^{-1}]

r_b Bubble radius   [m]

Sk Stokes number defined in Equation 3.1   [–]

U_S Slip velocity at the pipe wall   [m.s^{-1}]

x Length, measured positive upward, or in horizontal direction   [m]

Greek

$\dot{\gamma}$ Strain rate   [s^{-1}]

$\dot{\gamma}_W$ Strain rate at the pipe wall   [s^{-1}]

ε Volumetric liquid fraction in the foam   [–]

ε^* Liquid fraction at the flooding condition   [–]

μ Interstitial liquid dynamic viscosity   [Pa.s]

Π Dimensionless number defined in Equation 3.4   []

ρ_f Interstitial liquid density   [kg.m^{-3}]

ρ_g Gas density   [kg.m^{-3}]

σ Equilibrium surface tension   [N.m^{-1}]

τ Shear stress   [N.m^{-2}]

τ_W Wall shear stress   [N.m^{-2}]

τ_y Yield stress   [N.m^{-2}]

Appendix 3A: Derivation of a Hypothesized One-Parameter Descriptor of Determining the Hydrodynamic Condition

Consider a gas–liquid foam with a planar distribution of bubble radii $f(r_b)$ contained within a vertical tube. The number fraction of bubbles with diameter r_b and $r_b + \delta r_b$ is $f(r_b).\delta r_b$. The area of the horizontal plane associated with this fraction of bubble radii is taken as $Br_b^2 f(r_b).\delta r_b$, where B is a geometrical factor that is assumed to be independent of r_b. By invoking Equation 3.6,

$$\delta j_d = \frac{\rho_f g r_b^2}{\mu} m\varepsilon^n B r_b^2 f(r_b).\delta r_b \tag{3A.1}$$

$$\frac{\rho_f g \bar{r}_b^2}{\mu} m\varepsilon^n B \bar{r}_b^2 = \int_0^\infty \frac{\rho_f g r_b^2}{\mu} m\varepsilon^n B r_b^2 f(r_b) dr_b \tag{3A.2}$$

which, if one assumes that the liquid fraction ε is locally constant throughout the foam, simplifies to

$$\bar{r}_b^4 = \int_0^\infty r_b^4 f(r_b) dr_b \tag{3A.3}$$

or

$$\bar{r}_b = \left[\int_0^\infty r_b^4 f(r_b) dr_b \right]^{1/4} \tag{3A.4}$$

However, to correct the planar sampling bias inherent in the description of $f(r_b)$, the method of de Vries (1972) is adopted to give the one-parameter descriptor in terms of the true size distribution $g(r_b)$:

$$\bar{r}_b = \left[\frac{\int_0^\infty r_b^5 g(r_b) dr_b}{\int_0^\infty r_b g(r_b) dr_b} \right]^{1/4} \tag{3A.5}$$

It is stressed that this one-parameter descriptor is a hypothesis only, and has not been tested experimentally.

References

Bernal, J.D., and Mason, J. 1960. Co-ordination of randomly packed spheres, *Nature*, 385, 910–911.

Bikerman J.J. 1938. The unit of foaminess, *Transactions of the Faraday Society*, 34, 634–638.

Boonyasuwat, S., Chavadej, S., Malakul, P., and Scamehorn, J.F. 2003. Anionic and cationic surfactant recovery from water using a multistage foam fractionator, *Chemical Engineering Journal*, 93, 241–252.

Boussinesq, J. 1913. On the existence of a superficial viscosity grade, in the thin layer of transition which separates a liquid from another adjacent fluid, *Annales de Chemie et de Physique*, 29, 349–357.

Briceño, M.I., and Joseph, D.D. 2003. Self-lubricated transport of aqueous foams in horizontal conduits, *International Journal of Multiphase Flow*, 29, 1817–1831.

Cheng, H.C., and Lemlich, R. 1983. Errors in the measurement of bubble-size distribution in foam, *Industrial & Engineering Chemistry Fundamentals*, 22, 105–109.

de Vries, A.J., 1972. Morphology, coalescence and size distribution of foam, bubbles, in *Adsorptive Bubble Separation Techniques*, Editor Robert Lemlich, Academic Press (New York), Library of Congress: 75-154398.

Deshpande, N.S., and Barigou, M. 2001. Foam flow phenomena in sudden expansions and contractions. *International Journal of Multiphase Flow*, 27, 1462–1477.

Durand, M., Martinoty, G., and Langevin, D. 1999. Liquid flow through aqueous foams: From the plateau border-dominated regime to the node-dominated regime, *Physical Review E*, 60, R6307–R6308.

Edwards, D.A., Brenner, H., and Wasan, D.T. 1991. *Interfacial Transport Processes and Rheology*, Butterworth-Heinemann.

Jastrzebski, Z.D. 1967. Entrance effects and wall effects in an extrusion rheometer during the flow of concentrated suspensions, *Industrial & Engineering Chemistry Fundamentals*, 6, 445–454.

Koehler, S.A., Hilgenfeldt, S., and Stone, H.A. 1999. Liquid flow through aqueous foams: The node-dominated foam drainage equation, *Physical Review Letters*, 82, 4232–4235.

Koehler, S.A., Hilgenfeldt, S., and Stone, H.A. 2000. A generalized view of foam drainage: Experiment and theory, *Langmuir*, 16, 6327–6341.

Kruglyakov, P.M., Karakashev, S.I., Nguyen, A.V., and Vilkova, N.G. 2008. Foam drainage, *Current Opinion in Colloid & Interface Science*, 13, 163–170.

Leonard, R.A., and Lemlich, R., 1965. A study of interstitial liquid flow in foam. I. Theoretical model and application to foam fractionation, *AIChE Journal*, 11, 18–25.

Li, X., Wang, X., Evans, G.M., and Stevenson, P. 2011. Foam flowing vertically upwards in pipes through contractions and expansions, *International Journal of Multiphase Flow*, 37, 802–811.

Martin, P.J., and Wilson, D.I. 2005. A critical assessment of the Jastrzebski interface condition for the capillary flow of pastes, foams and polymers, *Chemical Engineering Science*, 60, 493–502.

Mooney, M. 1931. Explicit formulas for slip and fluidity, *Journal of Rheology*, 2, 210–222.

Neethling, S.J., Lee, H.T., and Cilliers, J.J. 2002. A foam drainage equation generalized for all liquid contents, *Journal of Physics: Condensed Matter*, 14, 331–342.

Saint-Jalmes, A., and Langevin, D. 2002. Time evolution of aqueous foams: Drainage and coarsening, *Journal of Physics: Condensed Matter*, 14, 9397–9412.

Saleh, Z., Stanley, R., and Nigam, M. 2006. Extraction of polyphenolics from apple juice by foam fractionation, *International Journal of Food Engineering*, 2, Art. 2.

Shaw, R., Evans, G.M., and Stevenson, P. 2011. Start-up transients in a pneumatic foam, *Asia-Pacific Journal of Chemical Engineering*, 6, 613–623.

Stevenson, P. 2005. Remarks on the shear viscosity of surfaces stabilised with soluble surfactants, *Journal of Colloid & Interface Science*, 290, 603–606.

Stevenson, P. 2006. Dimensional analysis of foam drainage, *Chemical Engineering Science*, 61, 4503–4510.

Stevenson, P. 2007a. On forced drainage of foam, *Colloids & Surfaces A*, 305, 1–9.

Stevenson, P. 2007b. The hydrodynamic theory of rising foam, *Minerals Engineering*, 20, 282–289.

Stevenson, P., Ata, S., and Evans, G.M. 2007. Convective-dispersive transport of gangue through flotation froth, *Chemical Engineering Science*, 62, 5736–5744.

Stevenson, P., and Li, X. 2010. A viscous-inertial model of foam drainage, *Chemical Engineering Research & Design*, 88, 928–935.

Stevenson, P., and Li, X. 2012. Pneumatic foam, in *Foam Engineering:Fundamentals & Applications*, Editor Paul Stevenson, Wiley (Chichester), ISBN: 0470660805.

Stevenson, P., Mantle, M.D., Sederman, A.J., and Gladden, L.F. 2007. Quantitative measurements of liquid holdup and drainage in foam using NMRI, *AIChE Journal*, 53, 290–296.

Stevenson, P., Sederman, A.J., Mantle, M.D., Li, X., and Gladden, L.F. 2010. Measurement of bubble size distribution in a gas–liquid foam using Pulsed-Field Gradient Nuclear Magnetic Resonance, *Journal of Colloid & Interface Science*, 352, 114–120.

Wang, X., Evans, G.M., and Stevenson, P. 2013. Flooding in a vertically rising gas–liquid foam, Industrial & Engineering Chemistry Research.

Weaire, D. 2008. The rheology of foam, *Current Opinion in Colloid & Interface Science*, 13, 171–176.

Weaire, D., Pittet, N., Hutzler, S., and Pardal, D. 1993. Steady-state drainage of an aqueous foam, *Physical Review Letters*, 71, 2670–2673.

Yilmazer, U., and Kalyon, D. 1991. Dilatancy of concentrated suspensions with Newtonian matrices, *Polymer Composition*, 12, 226–232.

4

Mechanisms of Foam Instability

4.1 The Importance of Bubble Size to Foam Fractionation Performance

In Section 3.3 it was demonstrated that the hydrodynamic condition of a pneumatic foam is crucially dependent on the characteristic bubble size, and that changes in the size with height in the column can change the hydrodynamic performance (see Section 3.7). In particular, the liquid flux through the column is determined by the characteristic bubble size at the top of the column, which, if a froth launder is adopted, is the bubble size upon the free surface. Thus it is important to consider what the mechanisms of bubble size change (i.e., "disproportionation") are so that their impact on the hydrodynamic state can be understood. However, not only does the bubble size determine the hydrodynamics of foam fractionation column, but disproportionation of the foam as a function of height induces the mechanism of *internal reflux*, which will be discussed in Section 6.7.

Foam is an unstable system, and bubbles tend to enlarge, via the mechanisms of liquid film rupture and inter-bubble gas diffusion, in order to reduce the Gibbs free energy of the system. Consider a foam that has gas–liquid interfaces with surface tension (i.e., surface energy) of σ. The distribution of bubble radii is given by $f(r_b)$. It can be readily shown that if one assumes approximately spherical bubbles, the specific surface area S (i.e., the area of gas–liquid interface per unit volume of foam) is

$$S = \frac{3(1-\varepsilon)}{r_{32}} \tag{4.1}$$

where r_{32} is the Sauter mean of the bubble radius distribution, i.e.,

$$r_{32} = \frac{\displaystyle\int_0^\infty r_b^3 f(r_b)\,\mathrm{d}r_b}{\displaystyle\int_0^\infty r_b^2 f(r_b)\,\mathrm{d}r_b} \tag{4.2}$$

Thus, assuming that the surface tension is spatially constant, the surface free energy per unit volume of the foam is

$$G = \frac{3(1-\varepsilon)\sigma}{r_{32}} \tag{4.3}$$

Thus, the system can reduce the surface energy in the bulk of the foam by increases in the Sauter mean bubble radius, and this is achieved by inter-bubble gas diffusion and, to a lesser extent, liquid film rupture. However, film rupture can be a major destabilizing mechanism on the free surface of the foam.

4.2 Coarsening due to Inter-Bubble Gas Diffusion

Inter-bubble gas diffusion (also interchangeably known as Ostwald ripening, by analogy with the mass transfer process across grain boundaries of metals) of a gas–liquid foam is the process by which big bubbles consume adjacent smaller bubbles due to differences in pressure in the two bubbles caused by the Young-Laplace effect. The pressure within the smaller bubble is greater than in the larger one, and, if the diffusing gas is soluble in the liquid phase, this drives transport of the gas out of the small bubble and into the bigger one. The capillary pressure difference (i.e., the difference between the pressure in the gas and the adjacent liquid) for a generalized surface is

$$\Delta P = \sigma \left(\frac{1}{r_1} + \frac{1}{r_2} \right) \tag{4.4}$$

where σ is the surface tension and r_1 and r_2 are the principal radii of curvature. These are equal for the surface of a sphere; so for a spherical bubble with radius r, the capillary pressure difference is

$$\Delta P = \frac{2\sigma}{r} \tag{4.5}$$

The pressure in smaller bubbles is greater than in bigger bubbles, which is the driving force for mass transfer. This is shown schematically in Figure 4.1 for three bubbles that are spherical except for at the inter-bubble boundaries. The smallest bubble has a radius of curvature of r_3 but gas is drawn from here into the bubble of larger radius (and therefore lower pressure) r_2. Further, the biggest bubble has a radius of curvature of r_1 and is of the lowest pressure, and is therefore able to consume the bubble of intermediate size.

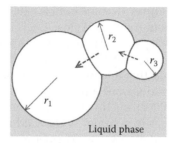

FIGURE 4.1
Schematic representation of three bubbles experiencing Ostwald ripening. The dashed arrows represent the direction of gas mass transfer.

Mass transfer would continue in the scenario represented in Figure 4.1 until the biggest bubble had consumed its two inferiors.

For a particular foam geometry, the rate at which inter-bubble gas diffusion proceeds is governed by

1. The solubility of the diffusing gas(es) in the interstitial liquid, and
2. The surface resistance to mass transfer caused by surfactant molecules at the gas–liquid interfaces.

The first effect of solubility on the rate of inter-bubble gas diffusion is best demonstrated by consideration of the relative stabilities of the foam "head" on the top of an Irish stout (e.g., Guinness) with that on the top of a lager-style beer. To some, it is common experience that a Guinness head is significantly more tenacious than that on the top of a lager, and the reasons for this are multifarious, but one of them is the different rates of Ostwald ripening. Guinness is typically dispensed using nitrogen gas, N_2 (or a mixture of gases that is predominantly nitrogen), whereas lager is dispensed under carbon dioxide (CO_2). However, the water solubility of CO_2 is much greater than that of N_2: The Henry's Law constant in water for CO_2 at 1 atm, 298K, and pH 7 is 0.035 mol.L^{-1}.atm^{-1} (Haynes, 2011) versus 0.00061 mol.L^{-1}.atm^{-1} for N_2 (Kavanaugh and Trussell, 1980). This means that the rate of inter-bubble gas diffusion for the CO_2 lager foam is greater, and the small bubbles are consumed at the expense of the larger ones, which creates an unstable foam.

The second factor in determining the rate of inter-bubble gas diffusion is the possibility of a surface resistance caused by adsorbed surfactant molecules that have been shown to be significant in some systems (Burnett and Himmelblau, 1970) but not in others (e.g., Hainwright et al., 2005).

There are two main routes to modeling Ostwald ripening. The first is due to Robert Lemlich (1978). Lemlich assumed that the surface resistance to mass transfer is negligible, but a slightly modified theory that allows for finite resistance has been given by Stevenson (2010). Lemlich's model enables the computation of the evolution of a bubble size distribution as gas diffusion proceeds. The model assumes that all bubbles in the foam are spherical, which is strictly true only in the Bernal close-packing limit (see Section 3.2). In fact, the adoption of a very wet foam in foam fractionation would result in a very poor enrichment ratio, and so significantly drier foams are typically utilized. We have observed that a foam of intermediate liquid fraction with polyhedral bubble shapes does not adhere to Lemlich's model of Ostwald ripening, and we posit that this is due to changes in the topology of a foam (rather than simply changes in volume) as Ostwald ripening proceeds.

The second route to modeling Ostwald ripening in foam has its genesis within the celebrated work of von Neumann (1952), who showed that a two-dimensional domain with n sides grows at a rate proportional to $n-6$. To put it another way, hexagonal domains stay the same size, pentagonal ones get smaller, and heptagonal domains grow. This theory has been extended to three dimensions (and beyond) by MacPherson and Srolovitz (2007). It is possible that a model for Ostwald ripening in foam could be based on this model, but to our knowledge, no theoretical work to do this has been proposed.

Because the mechanistic modeling of inter-bubble gas diffusion is problematical, three recent semi-empirical drainage laws are worthy of discussion. Vera and Durian (2002), by noting that the mean radius of bubbles in the scaling state, $<r>$, grows as the square root of time, gave the general expression that

$$\langle r \rangle \frac{d\langle r \rangle}{dt} = D^* F(\varepsilon) \tag{4.6}$$

where D^* is a "materials-dependent number with the units of a diffusion coefficient" and $F(\varepsilon)$ is a "universal function of only ε" (i.e., independent of $<r>$, the arithmetic mean bubble radius); and Stevenson (2010) showed that the former parameter is given by

$$D^* = \frac{\sigma RT}{P}\left[\frac{DHh}{2DH+\delta h}\right] \tag{4.7}$$

where σ is the equilibrium surface tension, T is the absolute temperature, R is the universal gas constant, D is the coefficient of molecular diffusion of the diffusing species in the liquid film, H is the Henry's Law constant of the diffusing species, δ is the thickness of the liquid film, h is the surface resistance to mass transfer, and P is the absolute pressure. (Note that the surface pressure is small compared to the absolute pressure, so the pressure P could be approximately the pressure within the liquid phase.)

There have been three functions suggested for $F(\varepsilon)$:

$$F(\varepsilon) = 1 - \sqrt{\frac{\varepsilon}{0.36}} \text{ (Bolton and Weaire, 1991)} \tag{4.8}$$

$$F(\varepsilon) = \left(1 - \sqrt{\frac{\varepsilon}{0.44}}\right)^2 \text{ (Hilgenfeldt et al., 2001a)} \qquad (4.9)$$

$$F(\varepsilon) = \frac{1}{\sqrt{\varepsilon}} \text{ (Vera and Durian, 2002)} \qquad (4.10)$$

It is claimed by the authors that Equation 4.10 correctly captures coarsening behavior for $0.03 < \varepsilon < 0.2$. Note that Equation 4.6, which can be derived via dimensional analysis, shows how the arithmetic mean bubble size changes, whereas in this is not the average defined in Equation 3.28, which is suggested for describing the hydrodynamic state of the foam, nor is it the Sauter mean bubble size (Equation 4.3) used for calculating enrichment (see Section 6.1).

The question remains as to whether Ostwald ripening is a significant mechanism of disproportionation. Stevenson et al. (2010) used Pulsed-Field Gradient Nuclear Magnetic Resonance (PFG-NMR) to measure the bubble size as a function of height in a non-overflowing pneumatic foam sparged with propane gas (which has a water-solubility that is intermediate between that of CO_2 and N_2). It was observed that there was no substantive change in bubble size distribution over a height of approximately 60 cm. This indicates that, at least in the system under investigation, Ostwald ripening (as well as liquid film rupture) did not have a material influence on the bubble size distribution. Further, a model based on Lemlich's method to describe Ostwald ripening, imperfect though it apparently is, predicts that the mechanism is insignificant.

This is, of course, not to say that there can never be a foam fractionation operation that is materially influenced by Ostwald ripening. However, not enough about Ostwald ripening in a pneumatic foam is known such that it can be relied upon to engender *internal reflux* if a very soluble gas such as CO_2 is used as the discontinuous phase, but if it is required to minimize disproportionation in a foam column as much as possible, then a very insoluble gas such as N_2 would be an appropriate choice.

4.3 Internal Bubble Coalescence

In addition to inter-bubble gas diffusion, the rupture of liquid films can change the bubble size distribution within the bulk of a pneumatic foam. However, less is known about the rupture of liquid films in the bulk of a foam than about films on a free surface of a foam (see Section 4.4), and the mechanisms that govern coalescence remain unclear (Colin, 2012). In the PFG-NMR experiments of Stevenson et al. (2010), there was no evidence whatsoever of internal bubble coalescence occurring, although this is certainly not the case in all foams.

The onset of internal coalescence is highly dependent on liquid fraction. Although the rupture of a single liquid film is a stochastic process, there is strong evidence that there exists a critical liquid fraction at which rupture commences. Carrier and Colin (2003) showed that, in a free-draining foam stabilized by a mixture of SDBS and $C_{10}E_{10}$, internal coalescence commenced when the liquid fraction reached a critical value of 0.0005 to 0.0007 (i.e., 0.05–0.07%), which is seen to be very low. The volume fraction threshold for the onset of bubble coalescence appears to be independent of bubble size. At values of liquid fraction above this, there is no demonstrable coalescence on a global scale. The critical liquid fraction is dependent on surfactant type and concentration. For instance, for foams stabilized by TTAB, the critical liquid fraction is observed to be approximately 0.0008 (0.08%) at the critical micelle concentration (CMC), but as low as 0.0002 (0.02%) when the concentration is ten times that of the CMC (Critical Micelle Concentration). The concentration of the adsorbing species in foam fractionation is typically lower than the CMC, and therefore one might suppose that the critical liquid fraction for internal coalescence would be still greater.

Operating a foam fractionation column at a liquid fraction of 0.08% is not beyond the bounds of reason for particularly high-value operations where high enrichment is desirable. However, such low liquid fractions would mean that the production rate is very low, and therefore higher liquid fractions are generally employed. Thus, for some foam fractionation processes, internal coalescence is likely to be insignificant, although for some, internal coalescence has been reported.

4.4 Film Rupture on the Free Surface of the Foam

In a foam fractionator that operates by discharging the foam into a launder vessel such that a free surface of foam is exhibited, surface bubble coalescence can have a significant impact on the hydrodynamic state of the foam—and therefore process performance. In much of the literature (for example, the model of the hydrodynamics of a froth flotation column of Neethling and Cilliers, 2003), there is no explicit distinction made between the physical mechanisms that control coalescence within the bulk of the foam and on a surface. However, Li et al. (2010) showed that the height achieved in a non-overflowing pneumatic foam (see Section 3.4) is dependent on the humidity gradient in the column. For instance, see the evolution of foam height in the column shown in Figure 4.2. Under certain conditions, and when the relative humidity at the top of the column was maintained

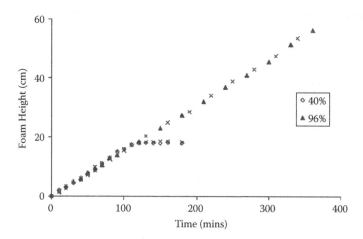

FIGURE 4.2

The behavior of a non-overflowing pneumatic foam is strongly dependent on environmental humidity, demonstrating that foam stability is governed by evaporation. (From Li, X., Shaw, R., and Stevenson, P. 2010. *International Journal of Mineral Processing*, 94, 14–19. With permission.)

at 40%, the foam grew as gas was sparged to it until an equilibrium height of somewhat less than 20 cm was established. However, when the same experiment was repeated while maintaining the relative humidity at the top of the column at 96%, the foam kept growing until it reached the top of the column. Further, Shaw et al. (2010) demonstrated that the liquid overflow rate into a launder, such as that found in some foam fractionation device designs, is dependent on the environmental humidity at the free surface.

Li et al. (2012) have speculated, with some preliminary supporting experimental data, that the cause of the significant dependency of free surface stability on environmental humidity is that nonuniform evaporation at the surface creates adverse Marangoni stresses that draw liquid from the top of bubbles and promote rupture.

There is no way, as yet, of predicting the extent of surface rupture of a foam but we can note the following qualitative observations:

1. A foam fractionator constructed with an inverted "U" at the top to collect the foam will not suffer any surface coalescence.

2. If a launder foam collection device is employed, the liquid flux through the column (and the extent of internal reflux) can be controlled by manipulating the humidity.

3. The application of dry air is a very efficient method of foam breaking. In particular, in the laboratory setting, a hair dryer has been particularly useful for foam breaking.

4.5 Measurement of Foam Stability

Thus far in this chapter, the Ostwald ripening, internal coalescence, and surface coalescence mechanisms of foam stability have been discussed, although the treatment of all three has been predominantly qualitative in nature. Foam drainage was discussed in Section 3.2 and can be considered an additional mechanism of instability, but drainage rates from a particular foam cannot be predicted in an a priori sense and must be described empirically. The physics that underpin the mechanisms of instability are complex, but this complexity is exacerbated by the interdependent nature of the mechanisms that is shown schematically in the figure of Hilgenfeldt et al. (2001b), and reproduced here in Figure 4.3. A higher rate of drainage means a drier foam, which promotes both coalescence and inter-bubble gas diffusion, and the bigger bubbles produced create more drainage *et sic in infinitum*. Thus, the designer of a foam fractionation device typically seeks some empirical description of foam stability.

However, there is no such thing as an intrinsic measure of "foam stability." There are several commonly adopted techniques that purport to measure "foam stability" but in fact are proxies for the measurement of one or another of the instability mechanisms, or a combination of several such mechanisms.

Perhaps the most common method of measuring foam stability is the Bikerman (1938) experiment. Gas is sparged to a vertical column (the bubble size is undefined in the method description) at a known gas superficicial velocity, j_g, and the equilibrium height of the "non-overflowing pneumatic foam,"

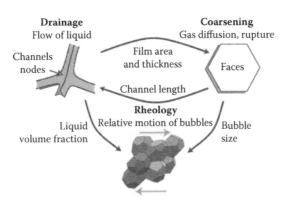

FIGURE 4.3
The schematic representation of the inter-dependency of mechanisms of foam instability. (From Hilgenfeldt, S., Koehler, S.A., and Stone, H.A. 2001. *Physical Review Letters*, 86, 4704–4707. With permission. Copyright (2001) by The American Physical Society. http://prl.aps.org/abstract/PRL/v86/i20/p4704_1).

l (see Section 3.4), is measured. Bikerman defined a so-called "unit of foam-ability" Σ; thus,

$$\Sigma = \frac{l}{j_g} \qquad (4.11)$$

which has units of reciprocal seconds. The Bikerman method of measuring foam stability is widely used across a range of process applications, but has found favor among foam fractionation researchers in particular. However, it is not a reliable method of measuring anything of practical interest to the designer of a foam fractionation device. Li et al. (2010) showed that the results of the Bikerman test were dependent on environmental humidity, and that almost any value of Σ that one cared to obtain was possible by humidity manipulation. To the best of our knowledge, not one previously published study that has invoked the Bikerman method has simultaneously controlled, or even measured, the humidity, so the measurements that are reported have no value. However, if the humidity is regulated, such as it was in the experiments of Li et al. (2010), then it is pertinent to contemplate what the dominant mechanisms of instability are in the Bikerman experiment: Certainly the surface rupture of bubbles makes a defining contribution to the outcome of the experiment but there may be nontrivial internal coalescence and, of course, drainage plays a role. Ostwald ripening is likely to be insignificant, as it was in the experiments of Stevenson et al. (2010). So, the Bikerman foam stability method is measuring a composite behavior of the foam that consists of surface rupture, foam drainage, and perhaps, to a lesser extent, internal coalescence. However, it does not measure an intrinsic value of "foam stability" because that it not defined.

The two methods of determining "foam stability" that are used in the brewing industry are the Rudi and the NIDEM tests. The details of these methods are given by Weaire and Hutzler (1999), but both depend on measuring the location of the liquid-foam interface as a function of time. The location of this interface is determined by coalescence and drainage, and it is stressed that the measurement of stability is a composite parameter that is a proxy for rates of different stability mechanisms. Further, the Ross and Miles (1941) test of foam stability, typically used for personal and domestic consumer products, involves pouring the solution onto a liquid surface and measuring the height of foam produced. Provision is made for wetting the walls of the column in order to reduce evaporation. Thus, it is clear that Ross and Miles anticipated a possible dependency of their test on humidity, but the top of the column is still open to the atmosphere. Again, the Ross and Miles test is a proxy for measuring a composite behavior of several instability mechanisms.

Other methods that measure the so-called "half-life" of foam have recently been suggested to provide a measure of foam stability. Farrokhpay (2011), among others, suggested measuring the time for one-half of the foam height in a tube to collapse as a measure of stability (which is largely a measure of surface instability), whereas Sun et al. (2013) have described stability by measuring the time it takes for one-half of the liquid to disengage from the foam in a tube (which is largely a measure of drainage rate).

Thus, the question arises as to what measure of foam stability is appropriate for the designer of a foam fractionation operation to adopt. Foam stability in foam fractionation determines the hydrodynamic state of the foam and therefore the fluxes of interstitial liquid and gas–liquid surface that report to the foamate. All the methods of estimating foam stability described above are of limited value in this context. In order to gain reliable information about how a particular foam will perform in a hydrodynamic sense during foam fractionation, it is probably necessary to conduct trial experiments under conditions that replicate the actual foam fractionation procedure. This is a part of the suggested design procedure discussed in Chapter 8.

4.6 Gas Expansion

Although not an instability phenomenon, changes in pressure in a foam fractionation column can influence bubble size as a function of height, and are therefore worthy of brief consideration in this chapter. The pressure at the bottom of the column is greater than at the top due to the weight of the liquid and gas phases. However, because foams tend to be relatively dry, this effect is almost always negligible. Consider an extreme example of a froth of height 5 m and liquid fraction of 33% (i.e., the wettest possible for a foam if $n = 2$; see Section 3.10). If the interstitial liquid has a density of 1,000 kg.m^{-3}, the mass of the gas is neglected, and the shear stress imparted by the pipe walls onto the foam is discounted, then the differential pressure between top and bottom is 3,240 Pa (or 3.2% of a standard atmosphere). So, if the top of the column is maintained at atmospheric pressure, the bottom would be at an absolute pressure of 104,565 Pa. Assuming that the discontinuous phase is an ideal gas, a bubble of 1 mm in diameter at the bottom would expand to a diameter of 1.010 mm at the top, which is not a difference worthy of consideration in practical design situations.

Nomenclature

Roman

D	Coefficient of diffusion of gas in the liquid solvent	[m^2.s^{-1}]
D^*	A quantity with dimensions of diffusion coefficient	[m^2.s^{-1}]

G	Specific surface energy	$[J.m^{-3}]$
h	Surface resistance to mass transfer	$[Pa.m^2.s.mol^{-1}]$
H	Henry's Law constant	$[mol.m^{-3}.Pa^{-1}]$
j_g	Gas superficial velocity	$[m.s^{-1}]$
l	Equilibrium height of foam in the Bikerman experiment	$[m]$
P	Pressure	$[N.m^{-2}]$
r	Radius of curvature	$[m]$
r_b	Bubble radius	$[m]$
r_{32}	Sauter mean bubble radius	$[m]$
$<r>$	Arithmetic mean bubble radius	$[m]$
R	Universal gas constant	$[J.K^{-1}.mol^{-1}]$
S	Specific surface area	$[m^{-1}]$
T	Absolute temperature	$[K]$

Greek

δ	Film thickness	$[m]$
ε	Liquid volume fraction	$[-]$
σ	Equilibrium surface tension	$[kg.s^{-2}]$
Σ	Bikerman's unit of foamability	$[s^{-1}]$

References

Bikerman, J.J. 1938. The unit of foaminess, *Transactions of the Faraday Society*, 34, 634–638.

Bolton, F., and Weaire, D. 1991. The effects of Plateau border in the 2-dimensional soap froth. 1. Decoration lemma and diffusion theorem. *Philosophical Magazine, B*, 63, 795–809.

Burnett, J.C., and Himmelblau, D.M. 1970. Effect of surface active agents on interphase mass transfer, *AIChE Journal*, 16, 185–193.

Carrier, V., and Colin, A. 2003. Coalescence in draining foams, *Langmuir*, 19, 4535–4538.

Colin, A. 2012. Coalescence in Foams, in *Foam Engineering: Fundamentals & Applications*, Editor Paul Stevenson, Wiley (Chichester), ISBN: 0470660805.

Farrokhpay, S. 2011. The significance of froth stability in mineral flotation: A review, *Advances in Colloid & Interface Science*, 166, 1–7.

Hainwright, J., Zhou, J., Evans, G.M., and Galvin, K.P. 2005. Influence of surfactant on gas bubble stability, *Langmuir*, 21, 4912–4920.

Haynes, W.M. 2011. *CRC Handbook of Chemistry and Physics, 92nd edition*, CRC Press, ISBN: 1439855110.

Hilgenfeldt, S., Kraynik, A.M., Koehler, S.A., and Stone, H.A. 2001a. An accurate von Neumann's law for three-dimensional foams, *Physical Review Letters*, 86, 2685–2688.

Hilgenfeldt, S., Koehler, S.A., and Stone, H.A. 2001b. Dynamics of coarsening foams: Accelerated and self-limiting drainage, *Physical Review Letters*, 86, 4704–4707.

Kavanaugh, M.C., and Trussell, R.R. 1980. Design of aeration towers to strip volatile contaminants from drinking water, *Journal of the American Water Works Association*, 72, 684–692.

Lemlich, R., 1978. Prediction of changes in bubble size distribution due to interbubble gas diffusion in foam, *Industrial & Engineering Chemistry Fundamentals*, 17, 89–93.

Li, X., Shaw, R., and Stevenson, P. 2010. Effect of humidity on dynamic foam stability, *International Journal of Mineral Processing*, 94, 14–19.

Li, X., Evans, G.M., and Stevenson, P. 2012. The effect of environmental humidity on static foam stability, *Langmuir*, 28, 4060–4068.

MacPherson, R.D., and Srolovitz, D.J. 2007. The von Neumann relation generalized to coarsening of three-dimensional microstructures, *Nature*, 446, 1053–1055.

Neethling, S.J., and Cilliers, J.J. 2003. Modelling flotation froths, *International Journal of Mineral Processing*, 72, 267–287.

Ross, J., and Miles, G. 1941. An apparatus for comparison of foaming properties of soaps and detergents, *Journal of the American Oil Chemists' Society*, 18, 99–102.

Shaw, R., Evans, G.M., and Stevenson, P. 2010. Dependency of liquid overflow rate upon humidity of a pneumatic foam, *International Journal of Mineral Processing*, 97, 26–30.

Stevenson, P. 2010. Inter-bubble gas diffusion in liquid foam, *Current Opinion in Colloid & Interface Science*, 15, 374–381.

Stevenson, P., Sederman, A.J., Mantle, M.D., Li, X., and Gladden, L.F. 2010. Measurement of bubble size distribution in a gas–liquid foam using Pulsed-Field Gradient Nuclear Magnetic Resonance, *Journal of Colloid & Interface Science*, 352, 114–120.

Sun, Q., Li Z., Stevenson, P., Jiang, L., Li, S., and Wang, J.Q. Particle-stabilised foam for effective displacement in porous media, *submitted to Journal of Petroleum Science and Engineering*.

Vera, M.J., and Durian, D.J. 2002. Enhanced drainage and coarsening in aqueous foams, *Physical Review Letters*, 88, 088304-1-4.

von Neumann, J. 1952. Discussion-Shape of metal grains, in *Metal Interfaces*. Editor Herring C. Cleveland, American Society for Metals, pp. 108–110.

Weaire, D., and Hutzler, S. 1999. *The Physics of Foam*, OUP, ISBN: 0198510977.

5

Hydrodynamics of Bubble Swarms

5.1 The Bubbly Liquid Layer in Foam Fractionation

As described in Section 1.2, a foam fractionation unit consists of a foam layer that is situated on top of a bubble liquid layer. The hydrodynamics of the foam layer were described in Chapter 3, with consideration for bubble stability in the foam layer given in Chapter 4. In this chapter the hydrodynamics of the bubbly liquid layer are considered.

Bubbles that are created at the bottom of the foam fractionation column rise through the liquid in what is known as a "swarm." Throughout the passage of a bubble through the liquid layer, surfactant species will adsorb onto the surface, and by varying the height of the bubble layer, the "residence time" of a bubble can be manipulated. This chapter begins by considering the behavior of a single gas bubble in an infinite liquid medium before adopting a description of the hindered settling of particles in slurries to estimate the hindered rising behavior of bubble swarms.

5.2 The Terminal Velocity of an Isolated Bubble

Before the hydrodynamics of a bubble swarm within a liquid is considered, we must first consider the behavior of an isolated single bubble rising in an infinite medium. Furthermore, we are going to assume, in the first instance, that the bubble is very small so that it is a sphere, and that the Reynolds number at the terminal rise velocity, V_∞, is low so that only viscous forces act on the bubble and inertial forces are negligible (i.e., the "creeping flow" assumption is valid). The Reynolds number is defined as

$$Re_\infty = \frac{\rho_f V_\infty d}{\mu} \tag{5.1}$$

and the assumption of creeping flow is good when Re_∞ is approximately less than 1. We define a dimensionless drag coefficient C_d such that

$$C_d = \frac{F_d}{\frac{1}{2}\rho_f V_\infty^2 A} \tag{5.2}$$

where F_d is the drag force imparted by the fluid on the bubble, and A is the projected area of the bubble (i.e., $\pi d^2/4$). Hadamard and Rybczynski (Rybczynski, 1911) calculated the drag force on spheres and from this work it can be shown that for a spherical bubble rising in a chemically pure liquid at low Reynolds number,

$$C_d = \frac{16}{Re_\infty} \tag{5.3}$$

At the terminal condition (i.e., the bubble is no longer accelerating), and if the weight of gas in the bubble is discounted, the drag is equal and opposite to the buoyancy which is given by

$$F_d = \frac{1}{2}\rho_f V_\infty^2 \frac{\pi d^2}{4} \frac{16}{Re_\infty} = 2\pi V_\infty d\mu = \frac{\pi d^3 \rho_f g}{6} \tag{5.4}$$

Equation 5.4 can be rearranged to give

$$V_\infty = \frac{\rho_f g d^2}{12\mu} \tag{5.5}$$

The reason for specifying that the liquid is chemically pure has to do with the boundary condition at the gas–liquid interface: If the liquid is chemically pure (i.e., there are no species present that can adsorb to the interface), then no shear stress can be supported at the interface and the drag is entirely due to variation in the pressure over the bubble surface (which is known as "form drag"). However, if the surface is entirely rigid, such as it is in a solid sphere, then the surface can support shear stress, which gives rise to "skin friction," which, along with the form drag, contributes to the total drag force. For a rigid sphere (Stokes, 1851) gave

$$C_d = \frac{24}{Re_\infty} \tag{5.6}$$

and thus, if a bubble's surface behaves like a rigid interface, we write

$$V_\infty = \frac{\rho_f g d^2}{18\mu} \tag{5.7}$$

By its very nature, the liquid in a foam fractionation column contains species that adsorb to the gas–liquid interface and so the interfaces are never completely fluid (i.e., they do not support a shear stress) but neither are they typically completely rigid. There exists an intermediate condition that was first calculated by Boussinesq (1913) and can be expressed as

$$C_d = \frac{8}{Re_\infty} \left(\frac{2+3\kappa}{1+\kappa} \right)$$

(5.8)

where κ is the surface dilatational viscosity, which determines the resistance of a surface to expansion (see Edwards et al. [1991] for a more rigorous definition). Unlike its cousin, the surface shear viscosity (i.e., the resistance to the shearing of a surface, which is important in theoretical treatments of foam drainage), the surface dilatational viscosity can be measured experimentally, for instance, using the oscillating pendant drop method employed by Cascão Pereira et al. (2003). Note that as $\kappa \to 0$, the Hadamard–Rybczynski solution (Equation 5.3) for a fluid sphere is recovered from Equation 5.8; and as $\kappa \to \infty$, the Stokes solution (Equation 5.6) for a rigid sphere is recovered.

However, isolated bubbles rising in infinite liquid media are not always spherical, and the bubbles are not always so small that the creeping flow assumption can be safely applied. At higher Reynolds numbers, the bubbles manifest an ellipsoidal shape with their major axis in the horizontal plane (i.e., normal to the direction of travel). For purposes of convenience, Clift et al. (2005) defined a spherical bubble as one that had a minor axis to major axis ratio (i.e., aspect ratio) of greater than 0.9, and defined as ellipsoidal bubbles as being more oblate. At still higher Reynolds number, these ellipsoids begin to "wobble" as they rise through the liquid.

Larger bubbles, or those in liquids of lower surface tension, can exhibit still different morphology. Clift et al. (2005) defined the shapes of "spherical-caps," "skirted bubbles," and "dimpled ellipsoidal-caps." The shape that the bubble adopts depends on the Reynolds number based on the diameter of the volume equivalent sphere, d_e:

$$Re_\infty = \frac{\rho_f V_\infty d_e}{\mu}$$

(5.9)

where

$$d_e = \left(\frac{6v}{\pi} \right)^{1/3}$$

(5.10)

and v is the volume of the bubble. In addition, the shape depends on the Eötvös number, *Eo*:

$$Eo = \frac{g(\rho_f - \rho_g)d_e^2}{\sigma} \qquad (5.11)$$

Clift et al. (2005) provided an approximate flow regime map that gave the bubble shape as a function of the Reynolds and Eötvös numbers, and the original plot is shown in Figure 5.1. In addition, the so-called "Eötvös plot" has contours of the Morton number, *M*, defined as

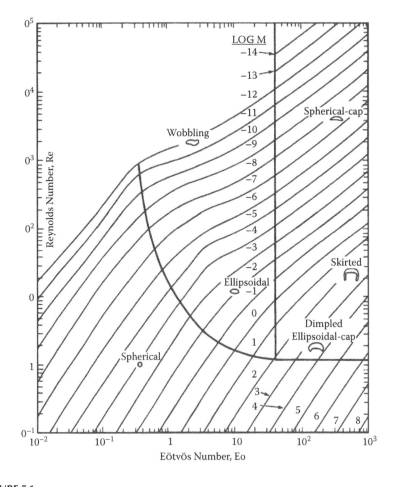

FIGURE 5.1
The "Eötvös plot" giving the shape and approximate terminal velocity of isolated bubbles in an infinite liquid medium. (From Clift, R., Grace, J.R., and Weber, M.E. 2005. *Bubble, Drops and Particles*, Dover, ISBN: 0486445801. With permission.)

$$M = \frac{g\mu^4(\rho_f - \rho_g)}{\rho_f^2\sigma^3} \tag{5.12}$$

which, in conjunction with a definition of a Eötvös number, allows the terminal velocity to be very approximately estimated via Re_∞. However, for a more reliable method of estimating the terminal velocity, across a range of parameters, Wallis (1974) correlated a dimensionless bubble radius:

$$r^* = r_b \left(\frac{g\rho_f^2}{\mu^2} \right)^{1/3} \tag{5.13}$$

with a dimensionless terminal velocity:

$$V^* = V_\infty \left(\frac{\mu g}{\rho_f} \right)^{-1/3} \tag{5.14}$$

A dimensionless group, as follows, is used to include a dependency on the surface tension:

$$P = \frac{\sigma^3 \rho_f}{g\mu^4} \tag{5.15}$$

Wallis suggested a number of dimensionless expressions based on P, r^*, V^* and the group:

$$B = \frac{g r_b^2 \rho_f}{\sigma} \tag{5.16}$$

For pure liquids (i.e., uncontaminated with surfactant) in the range $100 < P < 10^{10}$,

$$V^* = \sqrt{r^*} \qquad \text{for} \qquad B > 4$$

$$V^* = \sqrt{2}P^{1/12} \qquad \text{for} \qquad 1 < B < 4$$

$$V^* = \frac{r^{*2}}{3} \qquad \text{for} \qquad B < 1 \text{ and } r^* < 1.5$$

$$V^* = \min\left[0.408 r^{*1.5}, \sqrt{2} r^{*-1/2} P^{1/6} \right] \qquad \text{for} \qquad B < 1 \text{ and } r^* > 1.5 \tag{5.17}$$

For liquids contaminated with surfactant (such as those found in foam fractionation operations),

$$V^* = \sqrt{r^*} \qquad \text{for} \qquad B > 4$$

$$V^* = \sqrt{2}P^{1/12} \qquad \text{for} \qquad 1 < B < 4$$

$$V^* = \frac{2}{9}r^{*2} \qquad \text{for} \qquad B < 1 \text{ and } r^* < 1.5$$

$$V^* = \min\left[\sqrt{2}P^{1/12}, 0.307r^{*1.21}\right] \qquad \text{for} \quad B < 1 \text{ and } r^* > 1.5 \qquad (5.18)$$

It may appear curious as to why the Boussinesq solution for the general drag coefficient on a spherical bubble at low Reynolds number depends on the property of surface dilatational viscosity, whereas Wallis' generalized correlation has dependency on surface tension, and indeed the Eötvös plot has similar dependency. The explanation of this apparent dichotomy lies in the fact that Wallis' correlation and the Eötvös plot were developed by empirical observation over a wide range of physical parameters, whereas the Boussinesq solution is analytical and relevant to a very particular situation (i.e., creeping flow is assumed). Strictly speaking, Wallis should have observed the bubble velocities as a function of surface dilatational viscosity but he did not, and it is suggested herein that σ appears as a proxy for κ in his correlation.

Finally, it is stressed that only bubbles at the terminal condition have been thus far considered in this section. It is difficult to envisage a situation where one would need to calculate the acceleration of the bubble. However, if this is indeed required, it should be remembered that both the mass of the gas and the added mass must be accelerated. (At very low Reynolds numbers, the added mass is one-half of the mass of the displaced liquid.)

5.3 Hindered Settling of Particles in Slurries

Having developed methods for calculating the rise velocity of a single isolated bubble in an infinite liquid medium, it is necessary to consider how this velocity might be reduced because of the hindering effect of other bubbles and the container walls. However, there is no established method for calculating the extent of hindering for a bubble swarm; so instead, the method of Richardson and Zaki (1954) is described in this section, and then how it can be adapted to estimate the behavior of a bubble swarm is explained in Section 5.4.

Richardson and Zaki (1954) measured the hindered settling velocity of spherical particles of uniform diameter in a slurry, and, by dimensional analysis, showed that the ratio of the observed settling velocity to the terminal velocity of an isolated particle could be described as a function of the terminal Reynolds number, the volumetric liquid fraction, and the ratio of particle diameter to pipe diameter, that is,

$$\frac{V}{V_\infty} = f\left(Re_\infty, \varepsilon, \frac{d}{D}\right) \tag{5.19}$$

For convenience, we define $d/D = \lambda$. If the wall effects are discounted, Richardson and Zaki suggested that the velocity ratio was approximately a power-law function of the liquid fraction, with the power-law index, n, a function of the Reynolds number:

$$\frac{V}{V_\infty} = \varepsilon^n \tag{5.20}$$

Note that V is the velocity of particles measured relative to the stationary observer rather than relative to the liquid phase, which is a distinction that often causes confusion when this method is employed. The full Richardson and Zaki correlation for n is

$$
\begin{array}{llll}
n = 4.65 + 19.5\lambda & \text{for} & Re_\infty < 0.2 & \\
n = (4.35 + 17.5\lambda)Re_\infty^{-0.03} & \text{for} & 0.2 < Re_\infty < 1.0 & \\
n = (4.45 + 18\lambda)Re_\infty^{-0.1} & \text{for} & 1.0 < Re_\infty < 200 & \\
n = (4.45 + 18\lambda)Re_\infty^{-0.1} & \text{for} & 200 < Re_\infty < 500 & \\
n = 2.39 & \text{for} & 500 < Re_\infty &
\end{array} \tag{5.21}
$$

Although the above correlation is in widespread use in process industries, a single-expression correlation has subsequently been suggested by Khan and Richardson (1989); thus,

$$\frac{4.8 - n}{n - 2.4} = 0.043 Ar^{0.57}\left[1 - 2.4\lambda^{0.27}\right] \tag{5.22}$$

where the Archimedes number, Ar, is defined as

$$Ar = \frac{d^3 \rho_f (\rho_p - \rho_f)}{g\mu^3} \tag{5.23}$$

where ρ_p is the particle density. It should be stressed that the Richardson and Zaki correlation was developed for particles of uniform size. When the particle size becomes distributed, the hindered settling behavior becomes more complex (Greenspan and Ungarish, 1982), but this is beyond the current scope.

5.4 Hindered Rising of Bubbles

That bubbles in a swarm hinder the upward velocity of their peers is well known but there appear to be no dedicated and accessible methods to estimate this hindering behavior. Kay and Nedderman (1985) applied the method of Richardson and Zaki to describe the behavior of swarms, and their approach will be followed herein, with the proviso that it might not be accurate, for reasons described at the end of this section.

It can be assumed that, in typical foam fractionation devices, the ratio of bubble size to column diameter is so small that wall effects can be neglected (i.e., $\lambda \rightarrow 0$). Thus, if one calculates the terminal rise velocity of a single isolated bubble in an infinite liquid medium using the procedures described in Section 5.2, the hindered rise velocity can be calculated via the Richardson and Zaki correlation as a function of the liquid fraction. However, in order to estimate the liquid fraction in the bubbly liquid layer, we adopt the drift-flux method of Wallis (1969) that we now describe:

In order to quantify the hydrodynamic condition of the bubbly liquid layer, we must know the gas superficial velocity, j_g, and the liquid superficial velocity, j_f, in the column, noting that, at steady state, both fluxes are the same as they are in the foam layer. Wallis defined a slip velocity, V_S, between the absolute gas and liquid velocities, V_g and V_f, respectively, as follows:

$$V_S = V_g - V_f = \frac{j_g}{1-\varepsilon} - \frac{j_f}{\varepsilon} \tag{5.24}$$

The slip velocity can also be calculated by employing the Richardson and Zaki correlation; thus,

$$\frac{V_S}{V_\infty} = \varepsilon^{n-1} \tag{5.25}$$

Note that the index is now $(n-1)$ because the slip velocity is the velocity of the gas bubbles relative to the liquid phase, rather than measured relative to the stationary observer that gives rise to the index of n in Equation 5.20.

Eliminating V_S between Equations (5.24) and (5.25) gives

$$(1-\varepsilon)\varepsilon^n V_\infty = \varepsilon j_g - (1-\varepsilon) j_f \tag{5.26}$$

which can practically be solved numerically but will be solved graphically herein for illustrative utility. We define the "drift flux," j_{gf} as

$$j_{gf} = \varepsilon(1-\varepsilon)V_S \tag{5.27}$$

so that

$$j_{gf} = (1-\varepsilon)\varepsilon^n V_\infty \tag{5.28}$$

which is known as the "characteristic curve," and

$$j_{gf} = \varepsilon j_g - (1-\varepsilon)j_f \tag{5.29}$$

which is known as the "operating line." By plotting the drift-flux against liquid fraction using both Equations (5.28) and (5.29), the solution to the two simultaneous equations can be obtained. As will be seen in the worked example in Section 5.5, for upward concurrent flow (such as that typically seen in a foam fractionation device), there can be multiple solutions, but only one represents the condition of the bubbly liquid. Some workers have attributed the other solution to being representative of the foam layer but this is fallacious because of the presence of inter-bubble normal forces in the foam (see Li et al. [2012] for more details).

It should be noted that the single-expression correlation of Khan and Richardson (Equation 5.22) is probably less appropriate for application to bubble swarms because information about the terminal velocity is implicit within the correlation, and therefore a dependency on surface properties or bubble shape is not recognized.

Finally, the question arises as to whether the correlation of Richardson and Zaki is valid for application to bubble swarms. The answer is that the correlation probably does not give accurate estimates of the hindering effect of other bubbles on rise velocity. The reason for this assertion lies in fundamental differences in the fluid mechanics of settling slurries and rising bubble swarms. A settling particle exhibits a totally rigid boundary, whereas a bubble does not; and circulation of the discontinuous phase is possible in bubbly flow (i.e., circulation of gas within the bubble), whereas such circulation is not possible within a solid particle. It is for similar reasons that the effective viscosity of a low particle volume fraction slurry is given by Einstein (1906) as

$$\mu_e = (1 + 2.5\phi)\mu \tag{5.30}$$

(where μ_e is the effective viscosity of the colloid and ϕ is the volume fraction of the discontinuous phase), which is dissimilar to Taylor's (1932) analytical expression for the effective viscosity of a dispersed emulsion. Li and Stevenson (2013) have demonstrated preliminary experimental data that show that the use of a power-law relationship between the velocity ratio and liquid fraction does not always yield reliable results.

5.5 Worked Example: Calculating the Nature of Foam and Bubble Layers

Recall the worked example of Section 3.5 in which the hydrodynamic condition of a pneumatic foam was estimated. Under the conditions of $\mu = 1.0\,\text{mPa.s}$, $\rho_f = 1{,}000\,\text{kg.m}^{-3}$, $j_g = 8\,\text{mm.s}^{-1}$, $r_b = 0.5\,\text{mm}$, and adopting typical values of the foam drainage parameters, the liquid flux j_f was estimated at $0.533\,\text{mm.s}^{-1}$ and the liquid fraction a relatively wet 0.137. We seek to calculate the liquid fraction within the bubble swarm and the hindered bubble velocity. It will be assumed that the liquid feed is added to the very bottom of the column so that the liquid flux in the bubbly liquid is the same as it is in the foam.

We proceed by calculating the terminal rise velocity in an infinite liquid medium. To do this, we need the equilibrium surface tension (let us take, for the sake of argument, $\sigma = 0.04\,\text{N.m}^{-1}$). We note that the bubble diameter is 1 mm, and we assume that the bubbles are of uniform size. We will ignore gas compressibility effects on the size of the bubbles.

The dimensionless bubble radius r^* is calculated using Equation 5.13 as 10.70, the group P is 6.52×10^9 (Equation 5.15), and the group B is 0.0613 (Equation 5.16). The liquid is, of course, not chemically pure, so we select the appropriate expression from Equation 5.18 is

$$V^* = \min\left[\sqrt{2}P^{1/12}, 0.307r^{*^{+1.21}} \right]$$

The first term inside the square bracket computes to 9.30 and second term to 5.40, so the latter is selected as the minimum. Thus, the dimensionless velocity V^* is 5.40. Thus, Equation (5.14) gives a terminal velocity of $0.116\,\text{m.s}^{-1}$.

The terminal velocity can be estimated using the Eötvös plot (Figure 5.1). Assuming that the gas density is zero and using Equation 5.12, the Morton number is calculated as 1.53×10^{-10} (so that $\log M = -9.81$) and the Eötvös number (Equation 5.11) is 0.245. According to the Eötvös plot, the bubble shape is just within the spherical regime and the Reynolds number (Equation 5.9) is approximately 150, which would correspond to a terminal velocity of 0.15 m.s^{-1} which is the same order of magnitude as that calculated above using Wallis' generalized correlation. Wallis' result will be adopted throughout this worked example. The corresponding terminal Reynolds number is 116, which gives a Richardson and Zaki index of 2.77 via Equation 5.21 if wall effects are neglected.

The characteristic curve (Equation 5.28) and the operating line (Equation 5.29) are plotted for this example in Figure 5.2. It is seen that the relevant solution in the liquid fraction is 0.921 (i.e., 92.1%). There are two other solutions at around 7% and 22% but these are not physically relevant, and they certainly do not represent the condition of the foam layer.

It is reiterated that there is no experimental basis for the use of the Richardson and Zaki method for calculating the hindered rising of bubbles,

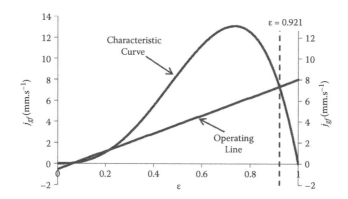

FIGURE 5.2
Graphical representation of Wallis' drift-flux method for the worked example in Section 5.5 demonstrating a liquid fraction in the bubbly liquid layer of 0.921 (92.1%).

and Kay and Nedderman (1985) stated that because the gas fraction rarely exceeded 0.2, that the simpler relationship

$$V^* = \min\left[\sqrt{2}P^{1/12}, 0.307r^{*1.21}\right] \tag{5.31}$$

could be substituted; that is, at low gas fractions, the slip velocity is approximately the same as the terminal velocity. This is equivalent to setting $n = 1$ in Equation 5.25. The approximate utility of Equation 5.31 can be seen by applying it to this worked example: The slip velocity is calculated by applying the method of Richardson and Zaki to be $0.116 \times 0.921^{2.77} = 0.093$ m.s^{-1} (i.e., 80% of the terminal velocity). Further, the use of Equation 5.31 has been supported by Li (2007), who measured the gas fraction $(1 - \varepsilon)$ as a function of gas superficial velocity, j_g, in a column with an internal diameter of 5 cm for the nisin solution that is enriched by foam fractionation as described in Chapter 10. He found that there was an approximately linear relationship as shown in Figure 5.3, which indicated that the value of n in Equation 5.25 is approximately unity at low gas fractions.

Nomenclature

Roman

A	Projected area of a bubble or particle	[m²]
Ar	Archimedes number defined in Equation 5.23	[–]
B	Dimensionless group defined in Equation 5.16	[–]
C_d	Drag coefficient defined in Equation 5.2	[–]

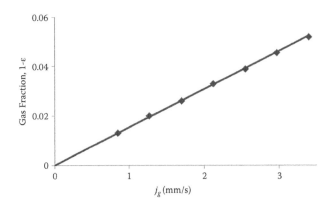

FIGURE 5.3
Gas fraction versus gas superficial velocity of bubbles through a nisin broth. The solid line represents $n = 1$ in Equation 5.25.

d	Bubble diameter ($\equiv 2r_b$)	[m]
d_e	Diameter of the volume equivalent sphere	[m]
D	Column diameter	[m]
Eo	Eötvös number defined in Equation 5.11	[–]
F_d	Drag force	[N]
g	Acceleration due to gravity	[m.s^{-2}]
j_f	Superficial liquid velocity	[m.s^{-1}]
j_g	Superficial gas velocity	[m.s^{-1}]
j_{gf}	Drift flux velocity	[m.s^{-1}]
M	Morton number defined in Equation 5.12	[–]
n	Richardson & Zaki index correlated in Equation 5.21	[–]
P	Dimensionless group defined in Equation 5.15	[–]
r^*	Dimensionless bubble radius defined in Equation 5.13	[–]
r_b	Bubble radius	[m]
Re_∞	Terminal Reynolds number defined in Equation 5.1	[–]
v	Bubble volume	[m^3]
V	Hindered rise velocity of bubble in a stationary frame	[m.s^{-1}]
V^*	Dimensionless bubble velocity defined in Equation 5.14	[]
V_f	Absolute velocity of the liquid phase	[m.s^{-1}]
V_g	Absolute velocity of the gas phase	[m.s^{-1}]
V_S	Absolute slip velocity between liquid and gas phases	[m.s^{-1}]
V_∞	Terminal velocity of a single isolated bubble in an infinite medium	[–]

Greek

ε	Volumetric liquid fraction	[–]
ϕ	Volumetric solids fraction	[–]

κ   Surface dilatational viscosity   [Pa.s.m]
λ   Ratio of bubble/particle diameter to column diameter   [–]
μ   Liquid dynamic viscosity   [Pa.s]
μ_e Effective dynamic viscosity of a suspension   [Pa.s]
ρ_f Liquid density   [kg.m^{-3}]
ρ_g Gas density   [kg.m^{-3}]
ρ_p Particle density   [kg.m^{-3}]
σ   Equilibrium surface tension   [N.m^{-1}]

References

Boussinesq, J. 1913. Existence of a superficial viscosity in the thin transition layer separating one liquid from another contiguous fluid, *Comptes Rendus Hebdomadaires des Séances de l'Académie des Sciences*, 156, 983–989.

Cascão Pereira, L.G., Théodoly, O., Blanch H.W., and Radke, C.J. 2003. Dilatational rheology of BSA conformers at the air/water interface, *Langmuir*, 19, 2349–2356.

Clift, R., Grace, J.R., and Weber, M.E. 2005. *Bubble, Drops and Particles*, Dover, ISBN: 0486445801.

Edwards, D.A., Brenner, H., and Wasan, D.T. 1991. *Interfacial Transport Processes and Rheology*, Butterworth-Heinemann, ISBN: 0750691859.

Einstein, A. 1906. Eine neue bestimmung der molekuldimensionen, *Annalen der Physik*, 19, 289–296.

Greenspan, H.P., and Ungarish, M. 1982. On hindered settling of particles of different sizes, *International Journal of Multiphase Flow*, 8, 587–604.

Kay, J.M., and Nedderman, R.M. 1985. *Fluid Mechanics and Transfer Processes*, Cambridge University Press, ISBN: 0521303036.

Khan, A.R., and Richardson, J.F. 1989. Fluid-particle interactions and flow characteristics of fluidised beds and settling suspensions of fluidised particles, *Chemical Engineering Communications*, 78, 111–130.

Li, X. 2007. Development of a Multistage-Draining Foam Fractionator & Its Application in Concentrating Nisin from its Fermentation Broth, M.Eng. thesis, Hebei University of Technology.

Li, X., and Stevenson, P. On power-law expressions of hindered rising of bubbles in swarms, submitted *International Journal of Multiphase Flow*.

Richardson, J.F., and Zaki, W.N. 1954. Sedimentation and fluidization. Part I. *Transactions of the Institution of Chemical Engineers*, 32, 35–53.

Rybczynski, W. 1911. Über die fortschreitende bewegung einer flüssigen kugel in einem zähen medium, *Bulletin International de l'Academie des Sciences de Cracovie*, 10, 40–46.

Stokes, G.G. 1851. On the effect of the internal friction of fluids on the motion of pendulums, *Cambridge Philosophical Transactions*, 9, 8–106.

Taylor, G.I. 1932. The viscosity of a fluid containing small drops of another fluid, *Proceedings of the Royal Society A*, 138, 41–48.

Wallis, G.B. 1969. *One-Dimensional Two-Phase Flow*, McGraw-Hill, New York.

Wallis, G.B. 1974. The terminal speed of single drops or bubbles in an infinite medium, *International Journal of Multiphase Flow*, 1, 491–511.

6

Modes of Operation

6.1 Preamble

In Chapters 2 through 6, the fundamental physical principles that underpin foam fractionation were described. Foam fractionation works because the target species adsorbs to gas–liquid interfaces, and foam supports a lot of gas–liquid interface but only small amounts of liquid so that the concentration of the target species is enriched. The processes of adsorption to gas–liquid interfaces and the hydrodynamics of pneumatic foam and bubbly liquids have been described to the extent that can be utilized by the designer of foam fractionators. The understanding of the fundamental physics will now be used in order to draw mass balances around foam fractionation columns in various configurations to enable insight into the potential process performance of the unit operation.

Before commencing the development of these mass balances, it is appropriate to formally present the two major performance metrics of a foam fractionation device, which are the *enrichment* and the *recovery*.

The *enrichment*, E, of a species is defined as the ratio of the concentration of target species in the foamate product, C_p, and in the feed, C_0, that is,

$$E = \frac{C_p}{C_0} \tag{6.1}$$

In a semi-batch process (i.e., batch with respect to the liquid phase), the concentration of foamate product decreases over time as the liquid pool becomes depleted in target species, and the enrichment is defined as the average value taken over the whole campaign (i.e., the quotient of the final concentration of the foamate collected and the feed concentration). In a system with more than one adsorbing species, the enrichment must be defined with respect to a particular species. In the analyses presented in this chapter, it is assumed that there is only one adsorbing species.

The *recovery*, *R*, is defined as the ratio of the amount of target species reporting to the foamate product to the amount present in the feed. It can be expressed as follows:

$$R = \frac{C_p Q_p}{C_0 Q_0} \qquad (6.2)$$

where Q_0 is the volume of liquid delivered as feed and Q_p is the volume of recovered foamate product. Again, if there are multiple adsorbing species, the recovery with respect to a particular species must be defined.

In general, there exists a trade-off between high recovery and high enrichment; a product stream that is highly concentrated in target species is often obtained at the expense of low recovery, and vice versa. It will be seen that this limitation is described by invoking the hydrodynamics of a pneumatic foam column that were described in Chapter 3, although methods of partially overcoming the limitation will be described in Chapter 9.

6.2 Batch Operation

The very simplest method of performing a separation by foam fractionation is to charge some liquid containing an adsorbing species to the bottom of a column and then sparge gas bubbles to it so that a foam is formed above the liquid layer. Liquid can then drain from the foam to make it drier, before more liquid, or gas, is pumped to the bottom to drive the dry foam from the top of the column. This arrangement is shown schematically in Figure 6.1;

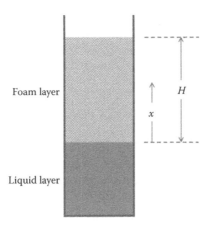

FIGURE 6.1
Schematic representation of the liquid and foam layers in a batch foam fractionation operation.

the height of the foam layer is H and the liquid fraction is dependent on the distance from the liquid-foam interface, $\varepsilon(x)$. The assumption is made that the interstitial liquid within the foam has the same concentration of adsorbing species as the concentration within the liquid layer. This is approximately valid because the action of molecular diffusion will tend to equilibrate the liquid concentration throughout the system. Further, it will be assumed that for every cycle of operation the recovery is low, so that the depletion of the liquid layer is insignificant, and the concentration stays approximately constant throughout the process at C_0. By assuming equilibrium between the concentration of the interstitial liquid and the surface excess via an adsorption isotherm we can write that

$$\Gamma = f(C_0) \tag{6.3}$$

and this surface excess is constant through the foam. (The reason for this assumption is that any concentration gradients in the interstitial liquid within the foam phase tend to be eradicated by molecular diffusion.) Thus, the amount of adsorbed species with the foam layer per unit cross-sectional area of the column is $\displaystyle\int_0^H S\Gamma\,dx$, where S is the specific surface area per unit volume of foam, and, if the expression of Equation 4.1 is substituted, gives

$$\Gamma \int_0^H \frac{3(1-\varepsilon(x))}{r_{32}(x)}\,dx \tag{6.4}$$

where r_{32} is the Sauter mean bubble radius. The volume of liquid within the foam per unit cross-sectional area of the column can be expressed as

$$Q_p = \int_0^H \varepsilon(x)\,dx \tag{6.5}$$

so the amount of target species dissolved in the interstitial liquid per unit cross-sectional area of the column is

$$C_0 \int_0^H \varepsilon(x)\,dx \tag{6.6}$$

Thus, the concentration of the foamate product is obtained by summing the amount of target species in both the interstitial liquid and adsorbed to the gas–liquid interfaces and dividing by the product volume, Q_p:

$$C_p = \frac{C_0 \displaystyle\int_0^H \varepsilon(x)\,dx + \Gamma \displaystyle\int_0^H \frac{3(1-\varepsilon(x))}{r_{32}}\,dx}{\displaystyle\int_0^H \varepsilon(x)\,dx} = C_0 + \frac{\Gamma \displaystyle\int_0^H \frac{3(1-\varepsilon(x))}{r_{32}(x)}\,dx}{\displaystyle\int_0^H \varepsilon(x)\,dx} \tag{6.7}$$

so that the enrichment (by Equation 6.1) is

$$
E = 1 + \frac{\Gamma \int_0^H \dfrac{3(1-\varepsilon(x))}{r_{32}(x)} dx}{C_0 \int_0^H \varepsilon(x) dx}
\tag{6.8}
$$

The recovery is given by the expression

$$
R = \frac{C_0 \int_0^H \varepsilon(x) dx + \Gamma \int_0^H \dfrac{3(1-\varepsilon(x))}{r_{32}(x)} dx}{C_0 Q_0}
\tag{6.9}
$$

The above expressions for enrichment and recovery are complicated by the fact that the bubble size and liquid fraction are allowed to vary as a function of the height up the column, x. A higher enrichment is obtained if the period for free drainage is increased. A treatment of free drainage is not included in this volume because batch operation is generally of only theoretical interest and would rarely be adopted in practice.

6.3 Semi-Batch Operation

The semi-batch mode of foam fractionation operation is best suited to separations in biotechnology where hygiene considerations preclude a fully continuous process. Liquid is charged batchwise to the column, and gas is sparged to the bottom such that foam rises up the column and is continuously collected at the top of the column. As the process proceeds, the liquid at the bottom of the column becomes depleted in target species. The foam fractionation ceases when either the concentration of surface-active material in the liquid is no longer able to support a foam capable of overflowing into the product foamate stream, or the instantaneous enrichment is deemed to have fallen to an unacceptably low level. Often, in practice (such as in the nisin process described in Chapter 10), a semi-batch operation is run until the residual liquid has fallen to a predefined level. After foaming, the remaining liquid in the column can potentially be recycled back to the unit operation, and the column cleaned before another batch is processed. The nisin production process described in Chapter 10 is operated in the semi-batch mode for reasons of biological hygiene, and also because the upstream fermentation is run in batch mode.

A semi-batch foam fractionation operation is shown schematically in Figure 6.2. Let a batch of liquid of volume Q_0 be charged to the bottom of the column at a concentration of adsorbing species of C_0. Gas is sparged to

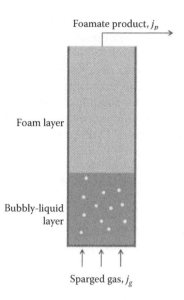

Foamate product, j_p

Foam layer

Bubbly-liquid layer

Sparged gas, j_g

FIGURE 6.2
Schematic representation of the bubbly liquid and foam layers in a semi-batch foam fractionation operation.

the bottom of the column at a superficial velocity of j_g so that the liquid flux (i.e., liquid superficial velocity) to the foamate product (the flux being based on the column cross-sectional area) is j_p. The process continues until a liquid volume of Q_1 (the residual volume) remains at a concentration of C_1. By a simple mass balance, the volume of foamate product recovered, Q_p, is

$$Q_p = Q_0 - Q_1 \tag{6.10}$$

and the concentration of the foamate product is

$$Cp = \frac{C_0 Q_0 - C_1 Q_1}{Q_0 - Q_1} \tag{6.11}$$

The enrichment is

$$E = \frac{Q_0 - \dfrac{C_1 Q_1}{C_0}}{Q_0 - Q_1} \tag{6.12}$$

and the recovery is

$$R = 1 - \frac{C_1 Q_1}{C_0 Q_0} \tag{6.13}$$

Unfortunately, Equations 6.10 through 6.13 cannot give quantitative estimates of the recovery and enrichment in the semi-batch process because no information about surface and liquid fluxes is given. Because the bulk concentration of the liquid layer decreases over time, so does the surface excess within the foam. This, in turn, impacts upon the drainage parameters, from Chapter 3, m and n (as the liquid depletes, the drainage rate tends to increase, causing a drier foam) and the foam stability (the foam becomes less stable giving bigger bubbles). Thus, as the foam fractionation proceeds, the instantaneous rate of recovery will decrease, whereas the instantaneous enrichment can either increase or decrease, depending on the extent of *internal reflux* (see Section 6.8) via the instability of the foam and the nature of the adsorption isotherm. However, it would be foolhardy to attempt an analysis of the process due to the complex and transient nature of the physical processes. A continuous process, such that all process variables are at steady state, is much more amenable to analysis, and this will be performed in the next section.

Although a priori design of semi-batch foam fractionation is difficult indeed, there is one valuable piece of information that should be taken experimentally. Given that the process only has utility if foam flows from the top of the column, it is of value to perform a Bikerman-style stability test (see Chapter 4) to ensure that overflow can occur at an acceptable level of liquid feed depletion.

6.4 Simple Mode Continuous Operation

Consider now a foam fractionation operation, shown schematically in Figure 6.3, in which liquid feed is continuously added to the bottom of the column with a continuous draw of spent liquid from the bottom pool, such that the foam is at "steady state." This configuration was referred to by Lemlich (1972) as "simple mode" foam fractionation. Such an arrangement is efficient in a process sense but its implementation in biological separations might be precluded due to hygiene considerations. The capital and running costs of a continuous process are greater than a semi-batch process and so, if regular equipment cleaning is essential, it is more practical to run the foam fractionator in semi-batch mode, with the batch time scheduled to coincide with the acceptable time between cleaning events, than it is to operate in continuous mode. However, for nonbiological foam separations, such as the stripping of detergents from laundry wastewaters or the stripping of frother in a froth flotation circuit, a continuous process may be desirable, specifically if the throughput of feed liquid is high or the upstream unit operation is also continuous.

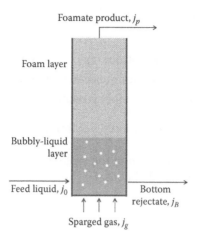

FIGURE 6.3
Schematic representation of simple mode continuous foam fractionation.

To analyze simple mode continuous foam fractionation, we begin by noting that the hydrodynamic condition of the rising foam can be calculated using the method given in Section 3.3. Given the liquid density and dynamic viscosity, the acceleration due to gravity, the bubble radius, the gas superficial velocity, and the drainage parameters m and n, the superficial liquid velocity to the product, j_p, can be calculated. The assumption is made that the bubbles are of constant size up the column and are approximately spherical so that the surface flux, j_S, is given by

$$j_S = \frac{3j_p}{r_{32}}$$ (6.14)

We will make three assumptions about the concentration equilibria within the system:

1. The bubbly liquid layer is "well mixed" so that the concentration of the bottom rejectate stream, C_B, is the same as the concentration of the liquid phase in the bubbly liquid, C.
2. The concentration of the interstitial liquid within the foam is the same as the concentration of the liquid phase in the bubbly liquid.
3. The surface excess, Γ, is uniform over the entire surface of the bubbles and is in equilibrium with the interstitial liquid.

There are seven further free variables in this problem:

1. The enrichment, E
2. The recovery, R

3. The feed concentration, C_0
4. The feed rate, j_0
5. The concentration of the bottom rejectate stream, C_B
6. The volume flux of the bottom rejectate stream, j_B
7. The concentration of the product stream, C_p

There are five independent equations that link these free variables:

1. An overall mass balance upon the system (assuming that the liquid phase is incompressible), that is,

$$j_0 = j_p + j_B \qquad (6.15)$$

2. A mass balance upon the adsorbed species:

$$j_0 C_0 = j_p C_p + j_B C_B \qquad (6.16)$$

3. An expression for the product concentration as a function of fluxes, and surface and bulk concentrations:

$$C_p = C_B + \Gamma \frac{j_S}{j_p} \qquad (6.17)$$

4. The definition of enrichment (i.e., Equation 6.1), which for a continuous system is re-written as

$$E = \frac{C_p}{C_0} \qquad (6.18)$$

5. The definition of recovery (i.e., Equation 6.2), which for a continuous system is re-written as

$$R = \frac{j_p C_p}{j_0 C_0} \qquad (6.19)$$

Thus, because there are seven variables and five independent equations, there is a well-posed problem if the values of two of the seven variables are defined. For example, if the feed rate and concentration are given (and if the hydrodynamic condition of the foam and the adsorption isotherm are known), then all other variables can be calculated through solution of the set of five simultaneous equations given above.

Worked Example:

It is required to enrich an aqueous solution of the commercial surfactant Triton X-100 by a factor in a foam fractionation column with feed flux of

2.00 mm.s^{-1} and feed concentration of 0.05 mol.m^{-3}. The solution will be sparged with bubbles of uniform radius r_b = 0.5 mm at a gas superficial velocity j_g= of 5 mm.s^{-1}. Calculate the recovery R and enrichment E.

Assume that the drainage parameters m = 0.016 and n = 2, and the adsorption isotherm (Darton et al., 2004) is

$$\Gamma = \frac{3.677 \times 10^{-6} C}{C + 2.112 \times 10^{-3}} \tag{6.20}$$

Using the methods in Chapter 3, the liquid fraction is calculated to be 0.0744 (i.e., 7.44%), liquid flux j_f is 0.185 mm.s^{-1}, and, by Equation (6.14), j_S is 30 s^{-1}.

By substituting the adsorption isotherm into Equation 6.17, we write

$$C_p = C_B + \frac{3.677 \times 10^{-6} C}{C_B + 2.112 \times 10^{-3}} \frac{j_S}{j_p} \tag{6.21}$$

By eliminating C_p from Equation 6.16 and solving simultaneously with Equation 6.15, we calculate that C_B = 0.00727 mol.m^{-3}, C_p = 0.470 mol.m^{-3}, and j_B is 1.82 mm.s^{-1}. By employing the definitions of recovery and enrichment, it is found that R = 0.868 (86.8%) and E = 9.40.

An important control variable is the feed flux to the foam fractionator. For the conditions described in this worked example, the feed concentration is maintained at 0.05 mol.m^{-3} and the feed rate allowed to vary; the resultant values of recovery and enrichment are plotted in Figure 6.4. It is seen that, at relatively low feed rates, the recovery is high but the enrichment is low. This is because the concentration of the liquid in the bubbly liquid is highly depleted and the surface excess is correspondingly low. As the feed rate increases, the concentration in the bubbly liquid increases too, which means

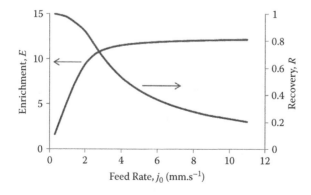

FIGURE 6.4
Curves of recovery and enrichment for the worked example in Section 6.4 using feed rate as the controlled variable.

that the enrichment is higher, but a higher concentration liquid is rejected from the bottoms and therefore recovery is lower. In the limit of infinite feed rate, the concentration of the bubbly liquid is the same as the feed and we write

$$\lim_{j_0 \to \infty} R = 0 \tag{6.22}$$

and

$$\lim_{j_0 \to \infty} E = 1 + \frac{j_s}{j_p C_0} \Gamma(C_0) \tag{6.23}$$

Equation 6.23 represents the maximum possible enrichment from a simple mode continuous foam fractionator.

6.5 Continuous Operation with External Reflux

Lemlich (1972) suggested improving the simple mode continuous foam fractionator described in Section 6.3 by taking some of the foamate product and returning it to the top of the column in a similar manner to that of washwater in flotation. The return of some of the foamate to the top of the column is known as "external reflux," as shown schematically in Figure 6.5. It will be assumed that the bubble size distribution stays constant up the column; bubble disproportionation is responsible for "internal reflux," which will be discussed in Section 6.7.

The hydrodynamics of such an operation are described in Section 3.3 but the features pertinent to external reflux are now reiterated:

1. The addition of an external reflux stream, at a superficial rate of j_R (with respect to the column cross-sectional area), causes the foam to become wetter (i.e., ε increases).
2. The addition of an external reflux stream reduces the net liquid flux, j_f, up the column within the foam.
3. No matter how much reflux is added to the foam, the flux in the overhead stream, previously called j_p but now designated j_T because the flux of liquid leaving the top of the column is not equivalent to the product stream flux, remains unchanged.

It will be recalled from Section 6.3 that it was assumed that the interstitial liquid within the foam had the same concentration as the bubbly liquid, and that the surface excess of the foam bubbles is in equilibrium with the liquid concentration. As such, it can be said that there is one "equilibrium stage"

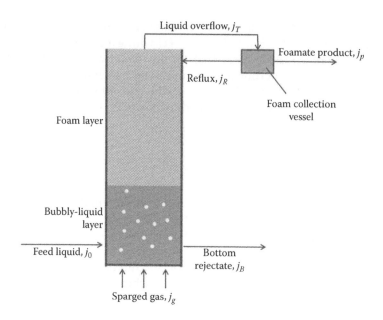

FIGURE 6.5
Schematic representation of a foam fractionation column undergoing external reflux.

(or "theoretical unit") within the column (i.e., the bubbly liquid and foam layer combined constitutes one equilibrium stage). However, if some of the enriched product is added to the top, in a manner reminiscent of distillation columns in which condensed overhead vapor is added as a reflux, then the surface of the bubbles contacts an interstitial liquid of higher concentration and this increases the surface excess. Thus, a higher enrichment is manifest. In fact, the addition of reflux potentially enables the establishment of multiple theoretical units. Mass balances upon such a column were first given by Stevenson and Jameson (2007) and are now reiterated with some minor modification of nomenclature.

We define a theoretical unit within the foam as a section of the column of finite height (the height of the theoretical unit, HTU) at which the surface excess of the surface leaving the stage from the top is in equilibrium with the concentration of interstitial liquid leaving the bottom (the unit, or stage, is assumed to be well mixed). The HTU is dependent on the rate of adsorption of target species to the bubble surface; an instantaneous adsorption would result in an infinitesimally high theoretical unit. Unfortunately, it is difficult to offer any method of calculation of the HTU beyond that given by Lemlich (1972) himself, who stated that, "Under reasonably favorable operating conditions, an HTU on the order of several inches or better can be expected, at least with simple surfactants." Presumably, Lemlich's use of the term "simple surfactant" alludes to a molecule of relatively low molecular weight (and therefore relatively high coefficient of molecular diffusion and high rate of adsorption). It is therefore supposed that molecules with relatively higher molecular weights, such as proteins, would exhibit a higher HTU. Citing an unpublished doctoral dissertation, Lemlich (1972) stated that, for the foam fractionation of SDS, the minimum HTU was experimentally found to be 1 inch (i.e., 0.0254 m) when the "flow number" F was 0.2905, where, in the current nomenclature and by adopting the current understanding of dynamic foam hydrodynamics,

$$F = \sqrt{\frac{\chi^3}{1+\chi+\dfrac{j_g}{j_T}}} \tag{6.24}$$

where χ is the fraction of the collapsed foam collected in the overhead that is refluxed back to the top of the column. It should be noted that, for the worked examples given herein, a value of 0.2905 is only approximately achieved at conditions of total reflux (i.e., $\chi = 1$). It must be stated, however, that no method for reliably estimating the HTU exists, and the best that can be done is to follow Lemlich's advice that it is several inches, and recognize that HTU will increase with the molecular weight of the adsorbing species (because the adsorption rate is lower).

Consider a column of foam with Z theoretical units (i.e., equilibrium stages) in the foam layer. It is assumed that the bottom bubbly liquid also forms a theoretical unit (by analogy with the reboiler in a distillation column), which is designated the bottom stage, B. Thus, the column has $(Z + 1)$ theoretical units in total. The feed to be enriched is added to stage B, and the bottom rejectate stream is drawn from stage B, which is considered to be well mixed. The top stream is collapsed (in a unit analogous to the condenser associated with a distillation column) and a fraction χ of the collapsed foam is refluxed back to the top of the column, with the balance of the foamate reporting to the product stream. See Figure 6.6 for a schematic representation of the discrete theoretical units.

It will be recalled from Section 6.4 that for simple mode continuous foam fractionation, if the hydrodynamic state of the foam and adsorption isotherm is specified, then the problem was well-posed with the specification of a further two variables from the list of seven. With external reflux, the problem becomes well posed after the reflux fraction χ and number of theoretical units in the foam Z are additionally specified. We have designated j_T as the liquid superficial velocity in the overhead stream. The foam collected from the top is collapsed and a fraction χ is returned as a reflux stream to the top of the column at rate j_R:

$$j_R = \chi j_T \tag{6.25}$$

and the flux of liquid to the product (again based on the column cross-sectional area, of course) is

$$j_p = (1 - \chi)j_T \tag{6.26}$$

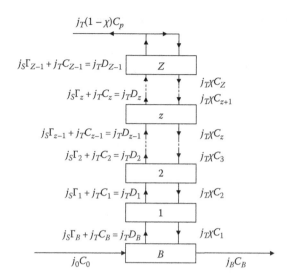

FIGURE 6.6
Discrete theoretical units in a foam fraction device operated with external reflux.

The overall mass balance now becomes

$$j_p = (1 - \chi)j_T + j_B \tag{6.27}$$

and the balance on the adsorbing species is

$$C_0 j_0 = C_B j_B + C_p(1 - \chi)j_T \tag{6.28}$$

The solution to the flow rates and concentrations in the streams is iterative. There are a number of solution strategies but if the feed condition is specified (i.e., j_0 and C_0 are given), the solution can proceed by allocation of a value of C_B followed by stage-wise mass balances before overall and component mass balances on the whole unit to assess the validity of the solution. A balance on the bubbly liquid (stage B) enables the calculation of the concentration of interstitial liquid in stage 1, C_1:

$$C_1 = \frac{j_S \Gamma_B + j_T C_B + j_B C_B - j_0 C_0}{\chi j_T} \tag{6.29}$$

which, by invoking the overall mass balance of Equation 6.27, can be rewritten as

$$C_1 = \frac{j_S \Gamma_B + (\chi j_T + j_0)C_B - j_0 C_0}{\chi j_T} \tag{6.30}$$

The balance around a subsequent general stage z gives:

$$C_{z+1} = \frac{j_S \Gamma_z + (1+\chi) j_T C_z - j_S \Gamma_{z-1} - j_T C_{z-1}}{\chi j_T} \qquad (6.31)$$

The product concentration is

$$C_p = C_Z + \frac{j_S}{j_T} \Gamma_Z \qquad (6.32)$$

which can be substituted into the component balance (Equation 6.28) to assess whether this balance is satisfied.

Worked Example:

Triton X-100 at a concentration of 0.05 mol.m^{-3} and a feed flux (with respect to the column cross-sectional area) of 2.0 mm.s^{-1} (i.e., the conditions in Section 6.3) is fed to a foam fractionation column with external reflux fraction $\chi = 0.5$ and two theoretical units in the foam phase (i.e., three theoretical units in total). The solution will be sparged with bubbles of uniform radius $r_b = 0.5$ mm at a gas superficial velocity $j_g = 5$ mm.s^{-1}. Assume the drainage parameters $m = 0.016$ and $n = 2$. Calculate the enrichment and recovery.

This is an iterative solution around the value of C_B. It is found that C_B takes the value of 0.00808 mol.m^{-3} and the concentration and surface excess of the two foam phase theoretical units are given in Table 6.1. The product concentration is 0.916 mol.m^{-3}, meaning the enrichment is 18.32 and the recovery is 0.846 (i.e., 84.6%). The product flux is 0.092 mm.s^{-1} (i.e., one-half of the product rate in the worked example of Section 6.4 because χ is 0.5).

Figure 6.7 shows the enrichment and recovery for the feed conditions of the worked example for one (solid line) and two (dashed line) equilibrium stages in the foam phase, as a function of reflux fraction. It is seen that enrichment increases monotonically with χ and recovery decreases. The maximum enrichment occurs under conditions of total reflux (i.e., χ is unity), taking the

TABLE 6.1

Interstitial Liquid Concentration, Collapsed Liquid Concentration, and Surface Excess of the Stages of the Worked Example in Section 6.5

Stage	C (mol.m^{-3})	D (mol.m^{-3})	Γ (mol.m^{-2}) × 10^{-6}
B	0.00808	0.48164	2.915
1	0.47420	0.61921	3.502
2	0.32257	0.91594	3.653

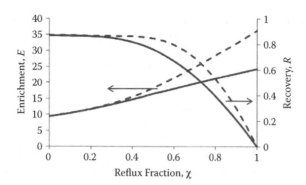

FIGURE 6.7
Enrichment and recovery for a column with one (solid lines) and two (dashed lines) theoretical units in the foam phase (i.e., two and three theoretical units in total).

value of 24.37 for the two-stage column and 36.29 for the three-stage column, but the recovery approaches zero (because there is no product flux). For both columns, the enrichment under zero reflux is 9.40, which is the same as in the simple mode continuous foam fractionation example of Section 6.4, and which would be expected because the condition of no reflux is equivalent to simple mode operation. In both cases, the maximum recovery is 87% which occurs under zero reflux and is the same value as in the simple mode example. Thus it is seen that reflux enhances enrichment but at the expense of recovery.

The fact that these solutions are iterative (and, it must be said, not particularly numerically robust) is unfortunate. Lemlich (1972) suggested the use of a graphical method that is analogous to the stage-wise solution for the operation of a binary distillation column (Smith, 1963), which proceeds by drawing "equilibrium" and "operating lines." For the foam fractionation column, Lemlich gave the correct equilibrium line but did not give the precise formulation for the operating line, and this deficiency is addressed herein. Lemlich defined a notional counter-current flow with a rising stream consisting of the gas–liquid interfaces with entrained liquid and a falling stream consisting of the reflux liquor. It is important to note that Lemlich did not suppose that this was the literal physical situation, and was simply employed as a device to enable a graphical mass balance just as it did in the stage-wise mass balance solution expounded above. The liquid flux of the rising stream is j_T, and each inter-stage rising stream can be considered a single phase liquid flow with the concentration of the collapsed foam, D, which is

$$D = C + \frac{j_S}{j_T} \Gamma(C) \tag{6.33}$$

This can be considered to give an equilibrium relationship between rising and falling streams leaving a stage, as the notional falling stream is of concentration C. In addition, by writing a mass balance around the bottom of the column that cuts the falling stream entering one stage and the rising stream leaving the same stage, we obtain the "operating line," which is

$$D = \chi C + \frac{j_0 C_0 - j_B C_B}{j_T} \tag{6.34}$$

which relates D leaving the top of one stage and C entering from the top of the same stage. It is noted that the operating line is linear because the upward and downward moving fluxes are constant. This is analogous to the constant molar overflow assumption in the graphical solution to the state of a distillation column that enables a linear operating relationship. The operating line has a slope that is equivalent to the reflux ratio; Lemlich (1972) recognized that the slope would be the ratio of notional upward and downward fluxes but did not equate this to the reflux fraction. Figure 6.8 shows both the equilibrium curve and the operating line for the above worked example with two stages in the foam phase. The bottom-left point is given by the equilibrium between C_B and D_B, and the solution is found by "stepping-off" alternately between the equilibrium curve and the operating line. The product concentration is the collapsed concentration of the stream leaving stage 2 (i.e., D_2).

It is unfortunate, however, that such a graphical representation does not overcome the need for an iterative solution, as the concentration of the bubbly liquid must still be determined for the stepping-off construction to proceed.

It is reiterated that, although there are rising and falling streams depicted in Figure 6.6, this is for purposes of the stage-wise mass balance and does not imply that the streams are physically separated. It is a device to enable solution of the performance of the column when, in fact, both interstitial liquid and gas–liquid interfaces are moving co-currently up the column. Martin et al. (2010) used the methods expounded in Chapter 3 to calculate

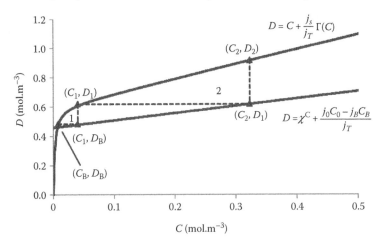

FIGURE 6.8
Graphical representation of the solution of the worked example in Section 6.5.

the hydrodynamic condition of a foam fractionation column undergoing total reflux (i.e., $\chi = 1.0$), and the theory was found to be effective in this respect. However, they assumed that the adsorption rate was instantaneous (the implication of which, in fact, would be that each theoretical unit would have an infinitesimal height) but defined a mass transfer coefficient between the apparently two counter-current streams. Although we disagree with the physical interpretation of Martin et al. (2010), their rigorous experiments on the enrichment of a stream of the surfactant cetylpyridiniumchloride (CPC) showed that the application of the hydrodynamic theory of rising foam was effective.

6.6 Stripping Mode Foam Fractionation

The foam fractionation operations described above have been for enriching the concentration of surface-active material in a liquid stream; the product is taken from the top, and the depleted liquid removed from the bottom and is either recycled upstream or discarded. However, such designs of foam fractionation are not necessarily the most appropriate when the objective is to remove unwanted surface-active material (for instance, in the removal of detergent from a laundry wastewater). Lemlich (1968) suggested a further mode of foam fractionation, "stripping mode," for applications stripping surface-active contaminants. As shown schematically in Figure 6.9, a contaminated stream is added to the top of the foam column (in a similar manner to that by which the reflux stream is added in Section 6.4) at a rate of j_0 and gas bubbles are sparged to the bottom at a superficial velocity of j_g.

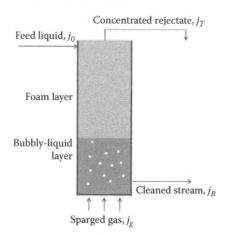

FIGURE 6.9
A schematic representation of Lemlich's (1968) stripping mode of operation.

The partially cleaned liquid is removed from the bottom of the column at a superficial velocity j_B (with respect to the column cross-sectional area) with the foam removed from the top of the column (at a rate j_T) being enriched in the unwanted contaminants. The advantages of such an arrangement for stripping operations are that effective reflux is engendered without having to take off, split, and return the foamate, and, depending on the height of the foam phase (and therefore number of theoretical units), has the potential to significantly deplete the concentration of adsorbing species.

A construction similar to that of Figure 6.8 can be drawn with the feed location at the top right of the diagram, but it should be noted that the operating line needs to change to reflect the different configuration of streams, so that

$$D = \chi C + C_T - \chi C_0 \qquad (6.35)$$

where the effective reflux fraction in this case is given by

$$\chi = \frac{j_0}{j_T} \qquad (6.36)$$

Distributing the feed to the top of foam column is practically challenging. Consider the application mentioned above, in which detergent is to be removed from the effluent of a commercial laundry. If the liquid feed is totally free of solids, then it might be appropriate to add the feed to the top of a perforated plate so that the liquid sprinkles through the holes. However, if solids are present, then these will tend to block the perforations and cause a reduction in performance. Therefore, it might be appropriate to include a solids removal operation upstream of the stripper.

Lemlich also suggested a "combined mode" of foam fractionation in which feed is added to the foam at a position intermediate between the top of the bubbly liquid and the top of the foam, along with a split of the foamate and addition as a reflux stream at the top of the column. This will engender a stripping section at the bottom of the column and an enriching section at the top. It is difficult to think of an application that would warrant such design complexity.

6.7 Internal Reflux

Throughout Section 6.5 the reflux was considered to be external, i.e., the foamate was collapsed and a fraction χ reintroduced to the top of the column. Implicit in the analysis was that the bubble size distribution was invariant all the way up the column. The hydrodynamic condition of the foam undergoing external reflux can be calculated by the method explained in Section 3.5: The locus of liquid flux versus liquid fraction remains the same, but because the addition of external reflux diminishes the net liquid flux and therefore the foam gets wetter. However, in Section 3.7 it was pointed out that an increase in the characteristic bubble radius serves the same purpose in that the increase of bubble size reduces the flux of liquid up the column, and the liquid that can no longer travel upward is liberated to move down the column just as it would if it were added as an external reflux stream. It was Lemlich (1968) who pointed out that bubble coalescence "destroys surface and so releases adsorbed material which flows back down through the rising foam. This rich drainage acts as internal reflux, which enriches the foam." Thus, when the reflux stream is added from the collapsed foamate stream, it is known as *external reflux*; but when it is formed from the foam partially collapsing within the column itself, it is known as *internal reflux*.

In fact, all rising pneumatic foams disproportionate (i.e., the bubble size distribution changes) with height to a greater or lesser extent. The amount of disproportionation in a column (and therefore the extent of internal reflux) is difficult to control and still more difficult to predict. However, it can be encouraged by sparging with a water-soluble gas such as CO_2 in order to promote inter-bubble gas diffusion. As will be explained further in Section 7.5, Winterburn and Martin (2009) proposed that the application of ultrasound to a foam can promote partial bubble collapse, and there is a potential application for its use in the control of internal reflux in foam fractionation, notwithstanding that it may cause protein denaturation. Another possible mode of control is to employ an open launder (see Section 7.4) as a collection vessel and manipulate the humidity of the air at the top of the column to control the rate of bursting on the free surface. (Note, of course, that bursting of bubbles of the free surface of a foam causes internal reflux just as disproportionation within the bulk does.) If internal reflux can be controlled, internal reflux would be preferred to external reflux because it does not require the splitting and return of some of the overhead stream to the top of the column, and therefore capital and operating costs are lower.

Stevenson et al. (2008) employed the hydrodynamic theory of foam (Section 3.3) to try to analyze a foam fractionation operation undergoing internal reflux. Of course, as already described, it is very difficult to even measure the bubble size distribution and still more difficult to predict how it changes up the column. However, what can be said is that, rather than the effective reflux flux j_R being constant throughout the foam later as it is when external reflux is applied, the amount of effective reflux decreases up the column. The foam at the very top of the column experiences no reflux at all because there are no bubbles above that are disproportionating. The very bottom of the foam layer experiences the full amount of reflux because all bubbles that are disproportionating reside above this level. As the size of the bubbles becomes larger, the amount of liquid that they can carry up the column diminishes, and this can be expressed by the following differential equation:

$$\frac{\partial j_R}{\partial r_b} = -\frac{\partial j_T}{\partial r_b}$$

(6.37)

Another way of expressing this is: if some function is known such that

$$j_T = f(r_b) \tag{6.38}$$

then

$$j_R = \int_{r_b^*}^{r_b} \frac{\partial f}{\partial r_b} dr_b \tag{6.39}$$

where r_b^* is the characteristic bubble radius at the very top of the foam.

6.8 Bubble Fractionation

In all the above analyses of foam fractionation, a central assumption has been that the bubbly liquid is well mixed (i.e., the concentration of the liquid is spatially uniform) and that adsorption equilibrium is attained therein. However, an allied technology to foam fractionation is that of "bubble fractionation," which was championed by Robert Lemlich (1968) and appears to have been first proposed in the public domain by Dorman and Lemlich (1965). The notion is that if the bubbly liquid is deep, then adsorbing species can be concentrated toward the top by the bubble swarm without the requirement of a foam layer. One application of such a method is in "continuous boosting" of a foam fractionation operation: If the feed is insufficiently rich in adsorbing species, then it may not be able to support a foam to enable foam fractionation to proceed. However, if the bubbly liquid can be enriched as a function of height in the column, then it may be that sufficiently large concentration is attained that can support a foam. Other than this, it is difficult to envisage a further application of bubble fractionation, as enrichment will always be greater if a foam is drawn from the top of the column (rather than a bubbly liquid).

Lemlich (1968) modeled the bubble fractionation process by assuming plug flow in both the liquid and the gas phases up the column. The adsorption process was modeled using the "penetration theory" first expounded by Higbie (1935), and expanded by Danckwerts (1951), which supposes that "parcels" of liquid are transported to the gas–liquid interface by turbulent eddies rather than a process of molecular diffusion. As explained by Stevenson (2012), in liquid–gas *absorptive* mass transfer operations, it is clear that penetration theory is more appropriate at the higher Reynolds number exhibited in a bubbly liquid than at the low Reynolds numbers exhibited in a foam column. Thus, its utility in *adsorptive* mass transfer in a bubbly liquid is certainly plausible. However, just as turbulent eddies promote mass transfer to gas–liquid interfaces, they can also promote axial mixing within the liquid phase of the bubbly liquid, thereby compromising the plug flow assumption.

Nonetheless, Shah and Lemlich (1970) showed that for a very deep bubbly liquid (concentration was measured over a height of 1.73 m) that a solution of crystal violet dye was approximately well-mixed just after bubbling commenced, but significant segregation occurred on the time scale of about one hour, when the concentration at the top of the column was four times the concentration towards the bottom. Thus it is apparent that, if a deep bubbly liquid is employed, the well-mixed approximation is not necessarily accurate, and its deficiency increases over time within a batch process. However, it is likely that, in a continuous operation, the addition of feed enhances mixing. In addition, concentration segregation within the bubbly liquid will enhance the performance of a foam fractionator, and therefore the well-mixed assumption should be retained in the interests of conservation design.

Nomenclature

Roman

C	Concentration of interstitial liquid	[mol.m^{-3}]
C_B	Concentration in the bottoms stream	[mol.m^{-3}]
C_p	Product concentration	[mol.m^{-3}]
C_T	Concentration in the overhead stream	[mol.m^{-3}]
C_z	Concentration leaving stage z	[mol.m^{-3}]
C_0	Feed concentration	[mol.m^{-3}]
C_1	Residual concentration in semi-batch mode	[mol.m^{-3}]
D	Collapsed foam concentration	[mol.m^{-3}]
D_z	Collapsed foam concentration leaving stage z	[mol.m^{-3}]
E	Enrichment factor	[–]
H	Height of the foam layer	[m]
j_B	Superficial liquid velocity of bottoms stream	[m.s^{-1}]
j_f	Superficial liquid velocity within the column	[m.s^{-1}]
j_g	Superficial gas velocity within the column	[m.s^{-1}]
j_R	Superficial liquid velocity of the reflux stream	[m.s^{-1}]
j_S	Surface flux	[s^{-1}]
j_T	Superficial liquid velocity of the overhead stream	[m.s^{-1}]
j_0	Superficial liquid velocity of the feed stream	[m.s^{-1}]
Q_p	Product volume	[m^3]
Q_0	Initial feed volume	[m^3]
Q_1	Residual volume in semi-batch mode	[m^3]

r_b	Characteristic uniform bubble radius	[m]
r_b^*	Characteristic bubble radius at the very top of the foam	[m]
r_{32}	Sauter mean bubble radius	[m]
R	Recovery	[-]
S	Specific surface area	[m^{-1}]
x	Vertical distance measured positive upwards	[m]
Z	Number of theoretical units in the foam phase	[–]

Greek

χ	Reflux ratio	[–]
ε	Volumetric liquid fraction	[–]
Γ	Surface excess	[$mol.m^{-2}$]
Γ_z	Surface excess leaving stage z	[$mol.m^{-2}$]

References

Danckwerts, P.V. 1951. Significance of liquid-film coefficients in gas absorption, *Industrial & Engineering Chemistry*, 43, 1460–1467.

Darton, R.C., Supino, S., and Sweeting, K.J. 2004. Development of a multistaged foam fractionation column, *Chemical Engineering & Processing*, 43, 477–482.

Dorman, D.C., and Lemlich, R. 1965. Separation of liquid mixtures by non-foaming bubble fractionation, *Nature*, 207, 145–146.

Higbie, R. 1935. The rate of absorption of a pure gas into a still liquid during short periods of exposure, *Transactions of the AIChE*, 35, 36–60.

Lemlich, R. 1968. Adsorptive bubble separation methods, *Industrial & Engineering Chemistry*, 60, 16–29.

Lemlich, R. 1972. *Adsorptive Bubble Separation Techniques* (Ed. R. Lemlich), Academic Press, Library of Congress Number 75-154398.

Martin, P.J., Dutton, H.M., Winterburn, J.B., Baker, S., and Russell, A.B. 2010. Foam fractionation with reflux, *Chemical Engineering Science*, 65, 3825–3835.

Shah, G.N., and Lemlich, R. 1970. Separation of dyes in nonfoaming adsorptive bubble columns, *Industrial & Engineering Chemistry Fundamentals*, 9, 350–355.

Smith, B.D. 1963. *Design of Equilibrium Stage Processes*, McGraw-Hill, New York.

Stevenson, P. 2012. Gas-liquid mass transfer in foam, in *Foam Engineering: Fundamentals & Applications* (Ed., P. Stevenson), Wiley, ISBN: 0470660805.

Stevenson, P., and Jameson, G.J. 2007. Modelling continuous foam fractionation with reflux, *Chemical Engineering & Processing*, 46, 1286–1291.

Stevenson, P., Li, X., and Evans, G.M. 2008. A mechanism for internal reflux in foam fractionation, *Biochemical Engineering Journal*, 39, 590–593.

Winterburn, J.B., and Martin, P.J. 2009. Mechanisms of ultrasound foam interactions, *Asia-Pacific Journal of Chemical Engineering*, 4, 184–190.

7

Bubble Production and Foamate Recovery

7.1 The Significance of Bubble Size

Bubble production in the process of foam fractionation is centrally important, not only because the whole process is based on the principle that surface-active materials adsorb to the surface of bubbles, but also because the bubble size determines the final efficiency of the process, in terms of enrichment and recovery. A smaller bubble size gives more interfacial area available for the target substance to adsorb. Smaller bubbles also rise more slowly in the liquid, allowing for more adsorption time (see Chapter 5). Thus, a smaller bubble size favors interfacial adsorption and gives a higher recovery rate. On the other hand, liquid drainage from a foam made of small bubbles is slower, resulting in a wet foam that contains a large amount of low-concentration interstitial liquid (see Chapter 3) that reduces the final foamate concentration, thereby compromising enrichment. Figure 7.1 is a simulation of the effect of bubble size on the enrichment and recovery of 0.5 g.L^{-1} SDS (sodium dodecyl sulfate) solution using the hydrodynamic theory of rising foam (Stevenson, 2007). Of course, many experimental studies show the same effect (e.g., Brown et al., 1999; Crofcheck and Gillette, 2003; Lee and Maa, 1986). It should be noted that if the bubble size increase is caused by an increase in gas flowrate, then the effect of superficial gas velocity on the enrichment and recovery of the process can override the effect of bubble size, as observed by Du et al. (2002). The effect of gas flowrate on bubble size distribution is discussed in the following sections.

Above we only discussed the general idea of "bubble size." In fact, unless the foam is generated under lab conditions from a single orifice with well-controlled gas pressure and gas flowrate, the size of the bubbles in the foam cannot be uniform. Narsimhan and Ruckenstein (1986) studied the effect of bubble size distribution on the enrichment of foam fractionation processes and found that the bubble size distribution affects foam drainage and stability and thus the final enrichment of the process.

In some other applications that involve gas bubbles, such as aerated fermentation, the smallest bubble size may give the best result (Bredwell and Worden, 1998; Kaster et al., 1990). In foam fractionation, as discussed,

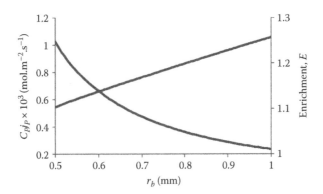

FIGURE 7.1
Enrichment and recovery as a function of bubble size for 0.5 g.L⁻¹ SDS solution sparged with
bubbles of radius 500 mm at 8 mm.s⁻¹. Drainage parameters of $m = 0.016$ and $n = 2$ are assumed.

the bubble size must be properly chosen for a specific application. A rule of
this is that for solutions with a low surfactant concentration and low foaming
tendency, such as a wastewater stream, a smaller bubble size normally gives
a better combination of enrichment and recovery. For a solution with high
surface-active material content such as a fermentation broth, using a larger
bubble size might be a good idea. To give the desired bubble size and bubble
size distribution, the bubble generation process must be properly engineered.
An appropriate bubble production method not only makes the process more
effective, but can also engender significant energy savings. Those commonly
used bubble generation methods are discussed below, as are some innovative
methods.

7.2 Sparger Types

Sparging (i.e., injecting gas into the foaming solution through a porous
medium) is probably the most commonly used method of bubble generation in foam fractionation and has been used, for example, in the studies of
Aksay and Mazza (2007), Crofcheck and Gillette (2003), Gerken et al. (2006),
Lee and Maa (1986), Noble et al. (1998), and Sarachat et al. (2010). The bubbles
then rise through the liquid pool as a bubbly liquid to form a foam layer.
The porous media used in foam fractionation are the same as in other bubbling applications such as distillation, fermentation, wastewater treatment,
and gas–liquid reaction processes. These can be sintered glass, metal frits or
pipes, perforated stainless steel plates or pipes.

Commercial products for sintered material spargers are readily available. The pore size of this type of sparger used in foam fractionation ranges

from several tens to several hundreds of microns (μm). Manufacturers of such spargers normally provide a nominal pore size; however, they rarely make any claim of the actual size of bubbles, because the bubble size varies depending on the physicochemical properties of the foaming liquid. Sizing of this type of sparger is based on the *superficial gas exit velocity* (the volumetric gas flowrate on the liquid side, divided by the sparger surface area) from the porous sparger surface. Generally speaking, the higher the superficial gas exit velocity, the larger the bubbles. There are special cases, such as at extremely low superficial gas exit velocity, where gas passes through only the larger pores of the sparger (if the pore size is distributed), so the average bubble size becomes larger at lower gas rates (Hoffer and Rubin, 1969). Such a situation is usually accompanied by liquid weeping into the gas line and should therefore be avoided. Thus, the total sparger surface area is designed to give a pressure difference across the sparger that is sufficient to prevent weeping for a given volumetric gas flowrate. The pressure drop for a given gas flowrate given by any specific sparger can generally be obtained from the manufacturer; such an example is given below for a typical range of spargers in Figure 7.2. The pressure drop depends on the thickness of the porous material, the pore size, and the pore gas velocity. For most applications using spargers with 10 to 100 μm pore size, a pressure drop of 20 to 70 kPa across the sparger is typical.

Unlike sintered spargers, perforated plates or pipes are generally custom-made. Larger holes (diameter > 1 mm) can be drilled, and smaller holes are more conveniently manufactured by laser-punching. Joshi's group (Kulkarni et al., 2009; Kulkarni and Joshi, 2011) provided some guidelines

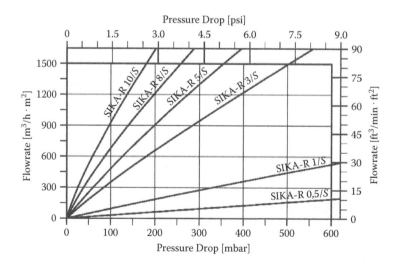

FIGURE 7.2
Typical pressure drop versus flowrate characteristics for a commercial range of spargers delivering air.

and empirical correlations for designing perforated spargers for bubble column reactors but the same principles should apply to foam fractionation as well because the liquid pool of a foam fractionation column is essentially a bubble column. The major challenge in designing this type of sparger is to ensure that the gas phase is evenly distributed across the cross-section of the column, and to prevent weeping of liquid. To overcome the potential for liquid weeping and blocking of pores when this type of sparger is used, the gas should be turned on before liquid is charged to the column, and then remain flowing until liquid is emptied at the end of the batch or campaign.

Sintered spargers are preferred if smaller bubbles are required (such as if the foaming solution has a very low surface-active material concentration and thus the foam is unstable). In the case of processing a fermentation broth with a high protein concentration, a relatively large bubble size may give a better combination of enrichment and recovery. For example, in the industrial production of nisin (see Chapter 10), stainless steel plates with laser-burned holes of around 135 μm were used as spargers. The bubbles generated from these spargers had a diameter of 3 to 5 mm.

Because bubble size is of critical importance in foam fractionation, it is a pity that there are no reliable a priori methods of determining the bubble size (or bubble size distribution) expected from a particular sparger. A rough estimate of bubble size as a function of the nominal pore radius r_0 can be obtained by calculating the largest bubble that can rest in equilibrium at the orifice of such a pore, which is given theoretically by Wallis (1969) as

$$r_b = \left[\frac{3\sigma r_0}{2g\left(\rho_f - \rho_g\right)} \right]^{1/3} \tag{7.1}$$

although Kutatelabze and Styrikovich, cited in Wallis (1969), suggested the following based on empirical observation:

$$r_b = 1.0 \left[\frac{\sigma r_0}{g\left(\rho_f - \rho_g\right)} \right]^{1/3} \tag{7.2}$$

As the gas flow rate increases, the bubble size increases above this value due to the finite time of detachment of the bubble from the orifice. However, as the gas flow rate increases still further, bubbles no longer form at the orifice, but instead gas passes as a continuous jet, which subsequently breaks into discrete bubbles as it passes into the liquid. Such jetting behavior creates bubbles that have a radius about twice the radius of the orifice, according to Wallis (1969).

7.3 Power Consumption of Bubble Sparging

Compression of the gas phase is required to make sparging possible. The gas must be compressed to overcome the pressure drop across the sparger, the hydrostatic pressure due to the liquid and frictional pressure drop due to the foam flow. Because foam fractionation is a very low-cost unit operation, the energy required by gas compression accounts for a major part of the total energy consumption. The power requirement for a reversible adiabatic (i.e., isentropic) compression of a perfect gas, $\dot{W}_{comp}$, is given by

$$\dot{W}_{comp} = Q_0 P_0 \frac{\gamma}{\gamma-1}\left[\left(\frac{P_1}{P_0}\right)^{\frac{\gamma-1}{\gamma}}-1\right] \tag{7.3}$$

where γ is the ratio of the isobaric and isochoric heat capacities; P_0 and Q_0 are the pressure and volumetric flowrate of the gas upstream of the compressor, respectively; and P_1 is the pressure of the compressed gas. In fact, practical compressors do not operate reversibly, and therefore it is necessary to define an isentropic efficiency, η_{is}, so that the power requirement is

$$\dot{W}_{comp} = \frac{Q_0 P_0}{\eta_{is}} \frac{\gamma}{\gamma-1}\left[\left(\frac{P_1}{P_0}\right)^{\frac{\gamma-1}{\gamma}}-1\right] \tag{7.4}$$

which can be reexpressed in terms of a mass flowrate $\dot{m}$.

$$\dot{W}_{comp} = \frac{\dot{m} R T_0}{\eta_{is}} \frac{\gamma}{\gamma-1}\left[\left(\frac{P_1}{P_0}\right)^{\frac{\gamma-1}{\gamma}}-1\right] \tag{7.5}$$

where T_0 is the absolute temperature of the gas upstream of the compressor and R is the specific gas constant. The isentropic efficiency of compressors typically lies in the range of 0.7 to 0.9 and is usually available from the manufacturer of the compressor. Values of R and γ for gases that might be selected for foam fractionation, including dry air, are given in Table 7.1. Although none of these gases are, in fact, perfect gases, Equation 7.5 is valid because the isobaric heat capacity does not change materially over the temperature range likely to be experienced by the gas as it passes through the compressor.

For example, to compress 50 kg.hr^{-1} (i.e., 0.0139 kg.s^{-1}) of dry air at 25°C and 1 atm to 2 atm (which is approximately required for a 5 m foam column with 30 kPa gauge hydrostatic pressure at the bottom, 50 kPa pressure loss through the sparger, and a 20 kPa pressure loss over the gas line and control

TABLE 7.1

Thermodynamic Properties of Gases That Potentially Have
Application in Foam Fractionation

	Dry Air	Nitrogen	Carbon Dioxide	Ozone
R [J.kg^{-1}.K^{-1}]	287	594	189	173
γ [-] at 25°C	1.40	1.30	1.30	1.29

valves) in a unit with isentropic efficiency 0.8 (i.e., 80%), the compressive
power requirement is 1.14 kW.

The energy required to generate the new gas–liquid interface can be calculated from the surface energy density of the bubbles, that is,

$$\dot{W}_{surf} = \sigma \dot{S} \tag{7.6}$$

where $\dot{S}$ is the rate of surface area generation and σ is the surface energy
density (or surface tension). The rate of surface area generation is given by

$$\dot{S} = \frac{3Q}{r_{32}} \tag{7.7}$$

where Q is the volumetric flowrate of gas and r_{32} is the Sauter mean bubble
radius defined in Equation 1.3.

The volumetric flowrate is a function of pressure, whereas the mass flowrate
is not. Consider a position at the bottom of the column in the above example
where the absolute pressure is 131.325 kPa, i.e., one standard atmosphere. At
a temperature of 25°C, this corresponds to a volumetric flowrate of 9.1 L. s^{-1}
(0.0091 m^3.s^{-1}). If the Sauter mean radius of the bubbles is 500 μm and the surface
tension is 0.04 N.m^{-1}, then the power required for production of a gas–liquid
surface is only 2.17 W (or about 0.2% of the power required for gas compression).

If the generation of a gas–liquid interface is the main purpose of sparging (the energy for the foam to rise through the column also originates
from the gas compressor), the energy efficiency of the process is seen to
be very low indeed. It can be seen that to improve the efficiency of sparging, the total pressure drop from the compressor to the bubbly liquid must
be reduced. In the context of sparger design, using a larger pore size and
pore number can significantly decrease the pressure drop over the sparger.
In fact, the pressure drop increases exponentially with a decrease in pore
diameter. However, with an increase in pore size, the bubble size inevitably increases, and therefore the specific surface area decreases. It would
be desirable to reduce the pressure drop of the sparger without sacrificing
the bubble size. Zimmerman et al. (2011) claimed to have (partly) solved
the problem using fluidic oscillation to generate bubbles of the same diameter as the orifice, but a full energy audit has not been carried out and

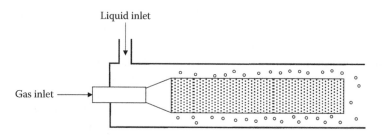

FIGURE 7.3
A schematic diagram of a possible arrangement of a dynamic sparger.

compressor power consumption has never been reported. It is expected that the compression requirement would be more or less the same as conventional sparging, and the only energy-saving part is in the actual bubble generation stage, which is negligible in comparison with the compressor energy consumption.

In fact, it has long been recognized that, in a stagnant liquid, the bubble size is much larger than the pore size where the bubbles come from. The situation can be improved significantly by applying a cross-flow of liquid at the surface of the sparger (marketed as "dynamic spargers" by many manufacturers). Figure 7.3 is an example of such a sparging system. A sintered sparger is encapsulated in a pipe. The gas is introduced into a sparger as usual, and a liquid flow is introduced into the annulus between the outside pipe and the sparger. Due to the shear stress imparted by the liquid upon the sparger surface, the bubble size thus generated can be much smaller than if the sparger is placed in a stagnant liquid. Note that the sparger can also be arranged so the liquid is inside and the gas flows in the annulus (Sadatomi, 2008). It is reported that in froth flotation, in situ static spargers have been being successfully replaced by dynamic spargers by Malhotra et al. (2009).

7.4 Other Methods of Bubble Production

While sparging is the most commonly used bubble generation method within the foam fractionation community, there are other bubble devices that are widely used in other processes such as flotation and aerated fermentation. The mechanism of these methods is discussed below and their potential application in foam fractionation is also discussed. Most of these methods focus on microbubble (with a diameter typically below 100 µ) generation.

Nucleation is a process that most people are familiar with because it creates the bubbles in carbonated beverages. When a beer bottle is opened, the pressure inside decreases, and the previously dissolved gas (in most cases, CO_2) comes out from the solution and forms the bubbles and a foam layer on the top. The same principle can be applied on an industrial scale to produce very small bubbles in a continuous fashion (Rodrigues and Rubio, 2007; Pérez-Garibay, 2012). This process involves the pre-saturation of process water with air at elevated pressure (3 to 6 atm), followed by a sudden reduction in pressure when the water is forced through a throttle (which can be needle valves or special orifices). Clouds of bubbles 20 to 100 µm in diameter are produced immediately after the release of pressure. An apparent disadvantage of this method is the high cost of water pre-saturation. The energy requirement for air compression can be calculated similarly to sparging, i.e., using Equation 7.3. The energy consumption can be reduced by adding a surfactant to the water before the saturation stage (Féris et al., 2000; Féris and Rubio, 1999; Takahashi et al., 1979), thereby reducing surface tension and therefore power consumption according to Equation 7.6. Rewriting Equation 7.6 for a single bubble of radius r_b, the minimum amount of energy, ΔE_{surf}, required for its formation is

$$\Delta E_{surf} = 4\pi r_b^2 \sigma \qquad (7.8)$$

It can be seen that the lower the surface tension, the lower the energy requirement for forming a bubble via nucleation. Féris and Rubio (1999) managed to reduce the saturation pressure from 3–6 atm to 2 atm using this method, but that is still significantly higher than the pressure differentials across most gas spargers. In addition to the high compression energy cost, bubble generation by dissolved air nucleation is typically not in situ. A separate saturator is required to pre-saturate the water, and an injection system is required to inject the microbubble-containing liquid into a column to form a foam. Gas can be introduced at the suction side of a positive displacement pump. Due to the high cost of air compression and the complexity of equipment requirements, application of this method in foam fractionation is not known to us. In addition, the bubbles generated by nucleation are not, in general, significantly smaller than the dynamic sparging process described above.

Bubble generation by mechanical agitation is another widely used technique, especially in fermentation (Kaster et al., 1990) as well as in flotation (Wills and Napier-Munn, 2005). Figure 7.4 shows a schematic diagram of a mechanical flotation cell. Turbulent flow is produced by the violent rotation of the impeller at the bottom of the cell. Air is delivered via the shaft of the impeller, and the stream is dispersed into small bubbles by the turbulent flow of the liquid phase and the relative velocity between the air and liquid phases. The bubbles rise to the top of the liquid and a froth layer is formed. The bubble size generated by this method is determined by the energy input and the interfacial tension. Generally speaking, the higher the impeller speed, the more the energy input, and, according to Equations

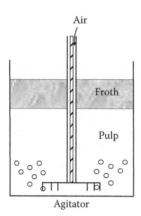

Air

Froth

Pulp

Agitator

FIGURE 7.4
A mechanical flotation cell. (Adapted from Wills, B.A., and Napier-Munn, T. 2005. Froth flotation, in *Wills' Mineral Processing Technology (seventh edition)*, Butterworth-Heinemann, Oxford.)

(7.6) and (7.7), the smaller the bubbles. To produce microbubbles in the 10 to 100 μm range, the impeller speed typically must be as high as 5,000 to 10,000 rpm. This represents very high energy intensity as the energy consumption of agitation is proportional to the cube of the agitation speed (McCabe et al., 2001). A reduction in gas–liquid surface tension can partially improve the energy efficiency of mechanical agitation because it has been shown by this method that a lower surface tension results in smaller bubbles (Shouci Lu and Eric, 2005). However, just as in the case of sparging, the energy requirement solely for creation of the gas–liquid interface is small in comparison to other sources of energy consumption. In the case of mechanical agitation, energy is dissipated to heat via Reynolds stresses caused by turbulence in the liquid phase. Agitation is generally considered an energy-intensive operation.

Mechanical agitation may be as energy intensive as dissolved air nucleation, but the equipment required is simpler, and attempts have been made to perform foam fractionation with it. Jauregi and Varley (1998) obtained very good results when creating microbubbles for protein recovery. At optimum conditions, the lysozyme recovery and enrichment ratio reached 9% and 19, respectively. In another study from the same group of researchers (Noble et al. 1998), the performance of a conventional sparging foam column and a mechanically agitated microbubble generator was compared with respect to protein recovery. The performance of the microbubble method, however, was not significantly better than conventional sparging. The reason for this was that foam consisting of microbubbles typically has a very high liquid fraction, as explained in Chapter 3. Because the concentration of the target substance in the liquid is low, the enrichment cannot be very high when a wet foam is employed.

Air entrainment by a moving liquid occurs naturally in the surf, in water-falls, and in mountain streams, and systematic studies on this phenomenon first appeared in the 1960s (Kenn and Zanker, 1967; Tong and Donnelly, 1966) and have included notable work by Bin (1993), Evans et al. (1996), and McCarthy (1972). Both laminar and turbulent jets entrain gas bubbles, although the exact mechanisms governing the entrainment processes of the two types are different (Tong and Donnelly, 1966; Ohl et al., 2000). The application of turbulent liquid jets for gas–liquid mixing and bubble genera-tion has been greatly promoted by the successful commercialization of the "Jameson Cell" in flotation (Simonis, 1981; Jameson, 1990, 1996), and some other applications such as in gas–liquid reactors (Concordia et al., 1987). Figure 7.5 shows the mechanism of gas entrainment in this type of device: An "induction trumpet" is formed at the point of jet impingement, and gas is drawn into the trumpet before being entrained into the receiving liquid and broken up into fine bubbles. If the ratio of the diameter of the liquid jet to the outside tube is properly designed, for example, 1:10 (Evans et al., 1996), intensive circulation can be engendered, resulting in high energy dissipation and the production of fine bubbles in the range of 100 to 500 μm diameter. In addition to the large interfacial area, the high turbulence generated by the jet also promotes the attachment of mineral particles to the bubble surface.

In plunging liquid jet bubble generation devices, the gas does not need to be compressed; this is because the gas is induced into the liquid due to the pressure reduction (Simonis, 1981; Jameson, 1996). From the discussion above we know that great energy savings can be obtained by eliminating the gas compressor. However, plunging jet devices require high jet velocity for the entrainment to occur, and Jameson (1996) has recommended a minimum jet velocity of 8 to 15 m.s^{-1}. In froth flotation, the plunging liquid jet is typically

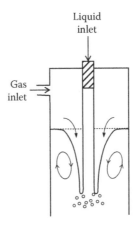

FIGURE 7.5
Schematic diagram for the gas entrainment mechanism of a liquid jet. (Adapted from Evans, G.M., Jameson, G.J., and Rielly, C.D. 1996. *Experimental Thermal and Fluid Science*, 12, 142–149.)

more energy efficient than mechanical agitation. In foam fractionation, whether the plunging liquid jet gives a net benefit over sparging, especially dynamic sparging, has yet to be determined with respect to foam fractionation performance.

From the above discussion it can be seen that, in order to make bubbles (i.e., to generate the gas–liquid interface), either the gas must be compressed or the liquid must be accelerated, or a combination of both, such as in dynamic sparging. If only larger bubbles (larger than 1 mm in diameter) are required, sparging might still be the most appropriate method, thanks to its flexibility and relatively simple configuration. Furthermore, dynamic sparging can generate much smaller bubbles at the cost of some additional piping and pump arrangement. The bubble size generated from liquid plunging falls in the same range as dynamic sparging but it is better suited for applications that require high intensity of mixing, such as fine particle flotation. None of the above-mentioned bubble generation methods were specifically designed for foam fractionation, and thus none of them take advantage of the properties of foam.

The fundamental reason that interfacial area generation is of low energy efficiency is that the gas phase is of low density and viscosity but high compressibility; therefore, it is difficult to increase the kinetic energy of a gas phase. Any method for bubble generation involving an increase in the kinetic energy of the liquid phase is more energy efficient than one involving an increase in the pressure of the gas phase. Although the gas fraction of foam can be as high as 98%, the foam is a much more rigid material than the gas itself. By exploiting this property of the foam, the authors of this monograph have invented a new concept of fine bubble generation, i.e., by liquid jet cutting (patent pending US61/757, 851).

Liquid jet cutting, more commonly known as "water jet cutting," is an industrial tool capable of cutting a wide variety of materials using a very high-pressure jet of liquid (mostly water). The concept of such cutting is not new, and a few patents can be found, dating back to the early twentieth century (Fourness and Pearson, 1935; Johnson, 1959). These early applications were at low pressure and restricted to soft materials such as paper and plastic. Modern high-pressure and high-velocity water jet cutters are capable of cutting metal and granite. However, foam is a very soft material, and the bubbles in a foam are fragile; therefore, even at very low pressure and low jet velocity (lower than the minimal jet velocity required for gas entrainment to occur in a plunging liquid jet device), larger bubbles in the foam can still be cut into smaller ones. Based on this principle, the authors of this book invented a microbubble generator that can make fine bubbles by cutting in the bubbles at the surface of an existing foam using multiple liquid jets. A schematic diagram of the device is shown in Figure 7.6. Large bubbles of 3 to 5 mm in diameter are generated by a coarse sparger; the bubbles are so big that the compressive duty to form them is low. These large bubbles rise through the liquid and form a thin layer of foam in the upper section of the column. Liquid is passed through a perforated plate and split into multiple liquid jets with a velocity

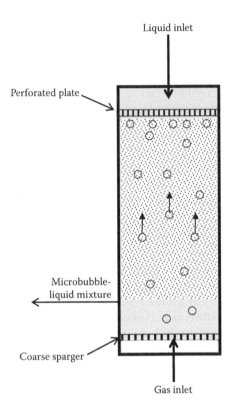

Liquid inlet

Perforated plate

Microbubble-
liquid mixture

Coarse sparger

Gas inlet

FIGURE 7.6
Schematic diagram of a new concept in microbubble generation.

of approximately 2 m.s^{-1}. The liquid jets impinging on the surface of the thin foam layer cut the large bubbles into smaller ones. When the bubbles become small enough and the buoyancy force acting on them becomes smaller than the drag force imparted by the down-flowing liquid, they are washed down the column, near the column wall, and exit from a port located in a lower section of the column in the form of a microbubble-liquid mixture, which is reminiscent of an emulsion. The gas fraction of this emulsion-like flow can be conveniently controlled by manipulating the volumetric flowrates of the liquid and gas flowrates. The device can operate stably at gas fractions up to 70%. In a continuous mode of operation, there is simultaneous upward and downward flow of the gas, which promotes better gas–liquid contact. A preliminary photographical measurement shows that the bubble diameter generated by this is below 60 μm at excellent energy efficiency.

The device shown in Figure 7.6 itself is not sufficient for foam fractionation operation because the product stream is a gas–liquid mixture rather than a dry foam as normally seen in a foam fractionation device. An external gas–liquid separation device must be used to separate the microbubbles from the

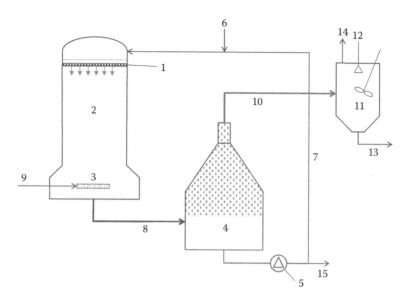

FIGURE 7.7
A foam fractionation process utilizing the microbubble generator: 1. Perforated plate; 2. Gas–liquid contacting column; 3. Coarse gas sparger; 4. Foam/liquid separator and foam collector; 5. Liquid pump; 6. Liquid feed; 7. Liquid with depleted surface-active materials; 8. Gas–liquid emulsion (liquid with high gas content); 9. Inert gas inlet; 10. Foam outlet; 11. De-foaming tank; 12. Anti-foam spray; 13. Foamate (product); 14. Gas outlet; 15. Tailings/rejectate (liquid).

emulsion. A full process flow diagram for a foam fractionation process that utilizes the microbubble generator is given in Figure 7.7. It is worth noting that the volumetric liquid flowrate required for cutting all the bubbles to a desired size might differ from that required for the feed flow rate. This can be overcome by simply mixing the fresh liquid feed with recycled liquid from the gas–liquid separator. Note that this does not increase the power consumption much because the liquid is flowing in a closed loop and thus there is no increase in hydrostatic head.

7.5 Foam Collection

Once adsorption has approached equilibrium, the foam is to be removed from the column and collapsed to give a concentrated liquid. Table 7.2 lists some typical arrangements for foam collection used by the foam fractionation community. Most of these methods involve a combination of contraction, expansion, and bending of the flow.

When the foam is dry, as it is in most cases of foam fractionation of protein when typically high enrichment is desired, the foam does not flow

TABLE 7.2

Foam Collecting Methods Used in Application-Orientated Foam Fractionation
Research

Researchers	System	Foam Collecting Method	
		Description	Schematic Diagram
Crofcheck and Gillette (2003)	Bovine serum albumin (BSA), water, nitrogen gas	Foam rises through the column, enters a narrowed pipe, and spills over the sides into an annular trough (foam collection cup)	
Brown et al. (1990)	Beta-casein, water, air	Column bends 180° downward at the upper end	
Jeong et al. (2003)	Protein from *Mimosa pudica L.*, seed, water, air	Top of the column is closed, and a side port is opened near the top of the column	
Saleh and Hossain (2001)	Mixture contains BSA, lactoferrin, water, air	Column bends 90° at the upper end, with an expansion	
Du et al. (2000)	Egg albumin and BSA, water, air	Foam overflows from the top of the column	Not given
Maruyama et al. (2007)	Egg albumin, water, nitrogen gas	Foam is aspirated out of the column	Not given
Burapatana et al. (2005)	Cellulase with the assistance of some detergents, water, air	Foam collection chamber is fitted to the top of the column	

under gravity due to its high effective viscosity. Because dry foam exhibits
Herschel–Buckley rheology (see the constitutive equation, Equation 3.20, in
Section 3.11), it cannot flow unless its yield stress is exceeded. Liquid drainage is minimal. In this case, the method of foam collection does not affect the
underlying performance of the process, but judicious selection of method can
improve the energy efficiency and convenience. As discussed at the beginning of this chapter, the driving force for foam flow originates in the gas
compressor; because the gas compressing system normally has low efficiency
with respect to surface creation, the pressure drop caused by the foam flow
may contribute significantly to the total energy consumption. In Table 7.2, all
but the first foam collection method are suitable for collecting dry foam.

The first arrangement in Table 7.2 is designed for collecting a wet foam and
is essentially the same as the froth launder used in froth flotation, as shown

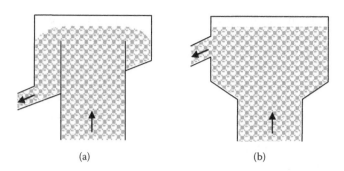

(a)   (b)

FIGURE 7.8
Illustration of: (a) a conventional froth launder; (b) a new concept in launder design.

in Figure 7.8(a). There is an expansion of flow area when the foam enters the launder, where the superficial gas velocity decreases. A certain amount of the interstitial liquid in the foam will drain out due to the expansion (as described in Section 3.9), resulting in a drier foam. Unfortunately, as the port is located at the bottom of the launder, the interstitial liquid still reports to the product stream, which decreases the enrichment of the process. On the other hand, in the launder design in Figure 7.8 (b), the liberated interstitial liquid flows back to the foam beneath; thus the foam leaving the system is drier, which enhances the enrichment. An additional benefit of the design in Figure 7.8(b) is that the interstitial liquid liberated from the foam can serve as a kind of internal reflux. Internal reflux is normally a consequence of bubble coarsening and rupture within the bulk foam (see Chapter 6; Lemlich, 1968; Stevenson et al., 2008), which enhances interface adsorption but has a negative effect on interfacial area. The new froth launder has the potential to generate internal reflux without sacrificing surface area, although the reflux stream will not be as enriched in adsorbing species as in conventional reflux. Therefore, the newly proposed froth launder design has the potential to reduce the volumetric flux and enhance interfacial adsorption at the same time, without impacting the surface area flux. All these factors favor the enrichment in the foam fractionation process.

7.6 Foam Collapse

The final stage of a foam fractionation process is the collapse of the foam. Whereas a relatively high foam stability is required during the foam fractionation process (although some instability is needed if internal reflux is desired), it immediately becomes undesirable as soon as the foam reaches the final stage. If no external reflux is used (as described in Section 6.5), a chemical antifoam (or defoamer) will be the most effective and cheapest choice (as it is in the nisin process described in Chapter 10). However, care must be taken to ensure that addition of the antifoam does not affect the functionality or quality of the final product. Slow mechanical agitation can help better distribute the antifoam and aids the process.

When external reflux is used, i.e., when part of the foamate is to be returned to the foam layer, the use of an antifoam agent should be avoided. This is because some antifoam used to collapse the foamate that contains antifoam would be returned to the column as part of the external reflux stream and therefore have deleterious effect upon the stability of the foam. Xu et al. (2010) found that the presence of antifoam compromised the effectiveness of foam fractionation of a protein solution by either reducing enrichment or recovery. Thus, under conditions of external reflux, it is necessary to implement nonchemical strategies of foam collapse. Darton et al. (2004) employed paddles mounted on a rotating shaft to collapse foam after each stage of their multistage foam fractionation column (described in Section 9.3), and reported apparent success. Winterburn and Martin (2009) employed low-powered (3 kW.m^{-3}) ultrasound at frequencies of 28 and 40 Hz at the outlet of a foam fractionator operating with the detergent Teepol™ as the adsorbing species. They did not find any direct evidence of enhanced foam collapse but did find that ultrasound enhanced foam drainage. (It is surmised that the mechanism by which this occurred is a reduction in surface shear viscosity at the gas–liquid interfaces.) Thus, it is inconclusive whether or not the application of ultrasound would be beneficial with respect to foam collapse on a commercial scale. In addition, it is not known whether the application of ultrasound would encourage the denaturation process in the foam fractionation of protein solutions.

We have found an effective way for manipulating the apparent stability of a foam: By manipulating the humidity of the air that is in contact with the foam (Li et al., 2012; Li et al., 2010) (i.e., at the free surface of the foam), the foam stability can be controlled. It has been observed that the foam collapses much faster under nonuniform evaporating conditions due to a surface-tension induced Marangoni effect. This method is very useful at the laboratory scale, where we use a standard hair dryer to increase evaporation and therefore collapse the foam. However, at the commercial scale, especially in the foam fractionation of biological molecules, it is envisaged that the preparation of a dry and clean gas stream might be prohibitively expensive.

Nomenclature

Roman

ΔE_{surf}	Surface energy of a single bubble	[J]
$\dot{m}$	Mass flowrate	[kg.s^{-1}]

P_0	Pressure upstream of the compressor	[Pa]
P_1	Pressure downstream of the compressor	[Pa]
Q	Volumetric gas flowrate in semi-batch mode	[$m^3.s^{-1}$]
Q_0	Volumetric gas flowrate upstream of the compressor	[$m^3.s^{-1}$]
r_b	Radius of a single bubble	[m]
r_0	Radius of an orifice	[m]
r_{32}	Sauter mean bubble radius	[m]
R	Specific gas constant	[$J.kg^{-1}.K^{-1}$]
$\dot{S}$	Rate of surface production	[$m^2.s^{-1}$]
T_0	Absolute temperature upstream of the compressor	[K]
$\dot{W}_{comp}$	Power required for gas compression	[W]
$\dot{W}_{surf}$	Power required for surface production	[W]

Greek

γ	Ratio of isobaric to isochoric heat capacities	[–]
η_{is}	Isentropic efficiency	[–]
ρ_f	Liquid density	[$kg.m^{-3}$]
ρ_g	Gas density	[$kg.m^{-3}$]
σ	Surface tension	[$N.m^{-1}$]

References

Aksay, S., and Mazza, G. 2007. Optimization of protein recovery by foam separation using response surface methodology, *Journal of Food Engineering*, 79, 598–606.

Bin, A.K. 1993. Gas entrainment by plunging liquid jets, *Chemical Engineering Science*, 48, 3585–3630.

Bredwell, M.D., and Worden, R.M. 1998. Mass-transfer properties of microbubbles. 1. Experimental studies, *Biotechnology Progress*, 14, 31–38.

Brown, A.K., Kaul, A., and Varley, J. 1999. Continuous foaming for protein recovery: Part I. Recovery of b-casein, *Biotechnology and Bioenineering*, 62, 278–290.

Brown, L., Narsimhan, G., and Wankat, P.C. 1990. Foam fractionation of globular proteins, *Biotechnology and Bioengineering*, 36, 947–959.

Burapatana, V., Booth, E.A., Prokop, A., and Tanner, R.D. 2005. Effect of buffer and pH on detergent-assisted foam fractionation of cellulase, *Industrial and Engineering Chemistry Research*, 44, 4968–4972.

Concordia, J.J., and Hall, D.R. 1987. Gas-Liquid Reactor and Method for Gas-Liquid Mixing, United States Patent US4683122.

Crofcheck, C., and Gillette, K. 2003. Evaluation of foam fractionation column scale-up for recovering bovine serum albumin, *Transactions of the ASAE*, 46, 1759–1764.

Darton, R.C., Supino, S., and Sweeting, K.J. 2004. Development of a multistaged foam fractionation column, *Chemical Engineering & Processing*, 43, 477–482.

Du, L., Loha, V., and Tanner, R.D. 2000. Modeling a protein foam fractionation process, *Applied Biochemistry and Biotechnology*, 84-86, 1087–1099.

Du, L., Prokop, A., and Tanner, R.D. 2002. Effect of bubble size on foam fractionation of ovalbumin, *Applied Biochemistry and Biotechnology*, 98-100, 1075–1090.

Evans, G.M., Jameson, G.J., and Rielly, C.D. 1996. Free jet expansion and gas entrainment characteristics of a plunging liquid jet, *Experimental Thermal and Fluid Science*, 12, 142–149.

Féris, L.A., Gallina, S.C.W., Rodrigues, R.T., and Rubio, J. 2000. Optimizing dissolved air flotation design system, *Brazilian Journal of Chemical Engineering*, 17, 549–556.

Féris, L.A., and Rubio, J. 1999 Dissolved air flotation (DAF) performance at low saturation pressures, *Filtration & Separation*, 36, 61–65.

Fourness, C.A., and Pearson, C.M. 1935. Paper Metering, Cutting and Reeling, United States Patent US2006499.

Gerken, B.M., Nicolai, A., Linke, D., Zorn, H., Berger, R.G., and Parlar, H. 2006. Effective enrichment and recovery of laccase C using continuous foam fractionation, *Separation and Purification Technology*, 49, 291–294.

GKN Sinter Metals (2013) Filter Elements catalogue.

Hoffer, M.S., and Rubin, E. 1969. Flow regimes of stable foams, *I&EC Fundamentals*, 8, 483–490.

Jameson, G.J. 1990. Column Flotation Method and Apparatus, United States Patent US4938865.

Jameson, G.J. 1996. Improved Column Flotation Method and Apparatus, Australian Patent AU9645592.

Jauregi, P., and Varley, J. 1998. Colloidal gas aphrons: A novel approach to protein recovery, *Biotechnology and Bioengineering*, 59, 471–481.

Jeong, G.T., Park, E.S., Wahlig, V.L. Burapatana, V., Park, D.H., and Tanner, R.D. 2003. Effect of pH on the foam fractionation of *Mimosa pudica L.* seed proteins. *Industrial and Engineering Chemistry Research*, 43, 422–427.

Johnson, C.O. 1959. Method for Cutting up Plastic and Semi-Plastic Masses, United States Patent US2881503.

Kaster, J.A., Michelsen, D.L., and Velander, W.H. 1990. Increased oxygen transfer in a yeast fermentation using a microbubble dispersion, *Applied Biochemistry and Biotechnology*, 24/25, 469–484.

Kenn, M.J., and Zanker, K.J. 1967. Aspects of similarity for air-entraining water flows, *Nature*, 213, 59–60.

Kulkarni, A.V., Badgandi, S.V., and Joshi, J.B. 2009. Design of ring and spider type spargers for bubble column reactor: Experimental measurements and CFD simulation of flow and weeping, *Chemical Engineering Research and Design*, 87, 1912–1630.

Kulkarni, A.V., and Joshi, J.B. 2011. Design and selection of sparger for bubble column reactor. Part II. Optimum sparger type and design, *Chemical Engineering Research and Design*, 89, 1986–1995.

Lee, J., and Maa, J.R. 1986. Separation of a surface active solute by foam fractionation, *International Communications in Heat and Mass Transfer*, 13, 465–473.

Lemlich, R. 1968. Adsorptive bubble separation methods: Foam fractionation and allied techniques, *Industrial and Engineering Chemistry*, 60, 16–29.

Li, X., Karakashev, S., Evans, G.M., and Stevenson, P. 2012. The Marangoni effect as a mechanism for the bursting of bubbles at the free surface of foam, *Langmuir*, 28, 4060–4068.

Li, X., Shaw, R., and Stevenson, P. 2010. Effect of humidity on dynamic foam stability, *International Journal of Mineral Processing*, 94, 14–19.

Lu, S., Pugh, R.J., and Eric, F., Editors 2005. Gas/liquid interfacial separation, in *Studies in Interface Science*, Elsevier, 559–645.

Malhotra, D., Taylor, P.R., Spiller, E., and LeVier, M. 2009. *Recent Advances in Mineral Processing Plant Design*, Society for Mining, Metallurgy & Exploration, Littleton, CO.

Maruyama, H., Seki, H., Suzuki, A., and Norio, I. 2007. Batch foam separation of a soluble protein, *Water Research*, 41, 710–718.

McCabe, W., Smith, J., and Harriott, P. 2001. *Unit Operations of Chemical Engineering, 7th edition*, McGraw-Hill.

McCarthy, M.J. 1972. Entrainment by Plunging Jets, Ph.D. thesis, University of Newcastle, Australia.

Narsimhan, G., and Ruckenstein, E. 1986. Effect of bubble size distribution on the enrichment and collapse in foams, *Langmuir*, 2, 494–508.

Noble, M., Brown, A., Jauregi, P., Kaul, A., and Varley, J. 1998. Protein recovery using gas–liquid dispersions, *Journal of Chromatography B: Biomedical Sciences & Applications*, 711, 31–43.

Ohl, C.D., Oguz, H.N., and Prosperetti, A. 2000. Mechanism of air entrainment by a disturbed liquid jet, *Physics of Fluids*, 12, 1710–1714.

Pérez-Garibay, R., Martínez-Ramos, E., and Rubio, J. 2012. Gas dispersion measurements in microbubble flotation systems, *Minerals Engineering*, 26, 34–40.

Porous metal design guidebook, www.pickpm.com/designcenter/porous.pdf, retrieved January 12, 2013.

Rodrigues, R.T., and Rubio, J. 2007. DAF–dissolved air flotation: Potential applications in the mining and mineral processing industry, *International Journal of Mineral Processing*, 82, 1–13.

Sadatomi, M. 2008. Micro-bubble Generator, Japanese Patent 4069211.

Saleh, Z.S., and Hossain, M.M. 2001. A study of the separation of proteins from multicomponent mixtures by a semi-batch foaming process, *Chemical Engineering and Processing*, 40, 371–378.

Sarachat, T., Pornsunthorntawee, O., Chavadej, S., and Rujiravanit, R. 2010. Purification and concentration of a rhamnolipid biosurfactant produced by *Pseudomonas aeruginosa* SP4 using foam fractionation, *Bioresource Technology*, 101, 324–330.

Simonis 1981. Method of and flotation cell for the flotation of solids. UK Patent Application GB 2107612A.

Stevenson, P. 2007. Hydrodynamic theory of rising foam, *Minerals Engineering*, 20, 282–289.

Stevenson, P., Li, X., and Evans, G.M. 2008. A mechanism for internal reflux in foam fractionation, *Biochemical Engineering Journal*, 39, 590–593.

Takahashi, T., Miyahara, T., and Mochizuki, H. 1979. Fundamental studies of dissolved air pressure flotation, *Journal of Chemical Engineering of Japan*, 12, 275–280.

Tong, J.L., and Donnelly, H.G. 1966. Gas bubble entrainment by plunging laminar liquid jets, *AIChE Journal*, 12, 563–571.

Wallis, G.B. 1969. *One-Dimensional Two-Phase Flow*, McGraw-Hill. Library of Congress Catalog Card Number 75-75170.

Wills, B.A., and Napier-Munn, T. 2005. Froth flotation, in *Wills' Mineral Processing Technology (seventh edition)*, Butterworth-Heinemann, Oxford.

Winterburn, J.B., and Martin, P.J. 2009. Mechanisms of ultrasound foam interactions, *Asia-Pacific Journal of Chemical Engineering*, 4, 184–190.

Xu, Z., Wu, Z., and Zhao, Y. 2010. Foam fractionation of protein with the presence of antifoam agent, *Separation Science & Technology*, 45, 2481–2488.

Zimmerman, W.B., Tesař, V., and Bandulasena, H.C.H. 2011. Towards energy efficient nanobubble generation with fluidic oscillation, *Current Opinion in Colloid & Interface Science*, 16, 350–356.

8

Column and Process Design

It might be astonishing to people outside the foam fractionation community that, although there are numerous studies showing the feasibility of the technique in various systems (see, for example, a recent review by Burghoff (2012)), there is little industrial uptake of the technology. However, froth-flotation, which is very similar to foam fractionation in many aspects, is widely used in the mining industry. The ultimate reason for this arises from the direct dependency of foam fractionation efficacy on the physicochemical properties of the feedstock to be foamed and, due to the vast complexity of the feed stocks from various different sources, a universalized, precise approach for systematic process design of foam fractionation is nearly impossible. From this point of view, froth flotation can be more readily standardized because the liquid properties can be well defined and do not vary much from one type of mineral to another. Not surprisingly, in some sectors of the industry where the liquid properties are more or less constant or a precise system control is not required, such as the removal of dissolved proteins from aquaculture (Timmons et al., 1995), not only is foam fractionation more commonly used, but there are even commercial foam fractionator products of various designs, taking the name of a "protein skimmer" (Wu, 2004). Therefore in this chapter, the provision of a universal set of rules for device and process design will not be attempted, as that is impractical. Instead, we initiate a discussion about what should be considered and how to obtain the process parameters if foam fractionation is to be designed for a specific system. This will also serve as a concluding text for all the individual aspects of the foam fractionation process discussed in the foregoing chapters. We will discuss these different aspects in an order that mirrors the design process of foam fractionation for industrial use, which means they do not necessarily follow the same order they appeared in the previous chapters.

8.1 Determination of Mode of Operation

The various modes of operation were discussed in Chapter 6. Typically, a decision must be made regarding the mode of operation ahead of all the other design parameters as all the latter are dependent on the former. Factors affecting the decision making include the following:

- *Feedstock supply.* Generally speaking, if the unit operation upstream foam fractionation is in batch mode (for instance, if it is a batch fermentation process), semi-batch (continuous with respect to gas and batch with respect to liquid) foam fractionation is the logical choice. While it is possible to adapt continuous foam fractionation to match the schedule of the batch mode fermentation, an intermediate liquid vessel is required to hold the solution. However, due to the fickle nature of the foam fractionation process, precise control is difficult. One could easily experience situations where a new batch of feedstock arrives but the old batch has not yet been completely processed, or situations where the feedstock from a previous batch has been used up but a new batch is not yet ready so the foam fractionation has to pause, which essentially forces the foam fractionation into a batch mode. Conversely, if the feedstock is supplied in continuous mode, such as a relatively constant stream of wastewater or from a continuous fermentation, a continuous foam fractionation is better suited. Continuous operation is also better suited for high throughput operations.

- *Column efficiency.* Continuous operation has the advantage of being able to run continuously; thus there is less downtime of the plant. However, higher utilization with respect to time does not always mean that the column is better utilized. On the contrary, continuous operation typically has lower resource (i.e., energy and space) efficiency. For example, when foam fractionation is used to recover a high added-value food additive or a pharmaceutical product from a biological feedstock, which many believe is where the future of foam fractionation lays (Lockwood et al., 1997; Schügerl, 2000; Burghoff, 2012), it is desirable that the rejectate from the process contains as little of the objective species as possible. In a batch or semi-batch mode of operation, the surface-active material concentration in the liquid depletes with time and the foam stability decreases. The process ends when no stable foam is produced from the liquid. In a continuous mode, the surface-active material in the column is constant with time; and due to the well-mixed characteristic of the liquid pool, the rejectate has the same surface-active material concentration as the liquid in the column. Therefore, the rejectate from a continuous process always must be higher than the end point of a batch mode operation; otherwise no stable foam is produced. This certainly is not ideal for high added-value product recovery. In addition, because the surface-active material concentration in the liquid is low, its concentration in the foamate is also low, as it is governed by the adsorption isotherm. An analog of this difference between batch and continuous mode of operation is the difference between a plug flow reactor (PFR) and a continuously stirred tank reactor

(CSTR), where the CSTR has a lower efficiency because it operates at the outlet conditions, while the PFR typically has a higher driving force. To proceed further with the analogy, just as CSTRs can be operated in series to approximate PFR performance, continuous foam fractionation columns can also be operated in series to overcome this disadvantage of continuous operation, which will be discussed in Chapter 9.

- *Simplicity of operation and reliability.* Foam fractionation columns are simple in terms of both construction and operation, but that does not necessarily eliminate the possibility of device failure. For example, fouling of, or liquid weeping through, spargers can occur, especially in wastewater treatment where the feedstock contains unknown amounts of suspended particles. Consideration must to be given to the consequences of such failures when making decisions on the mode of operation. It is easy to understand that continuous operation requires much higher reliability. As continuous operation runs for a prolonged time, reliable process monitoring and control are required, and this may cause practical difficulties in some systems (see Chapter 10 for an example).

- *Other considerations.* When foam fractionation is used to recover food or pharmaceutical products, the process is subjected to certain hygiene requirements. Because foam fractionation is typically operated at room temperature and the feedstocks from food or biological processes are rich in nutrients, sterilization and cleaning of the columns and pipe lines are necessary and may require the columns to be emptied on a regular basis; this renders the apparent advantages of continuous operation less attractive.

8.2 Column Dimensions

8.2.1 Diameter

It is noted that, throughout this book, gas and liquid flowrates are expressed as a superficial velocity (or flux) rather than as volumetric or mass flowrates. The fundamental reason for such an approach is that it enables a one-dimensional description of the process, which is appropriate because almost all flows in the foam layer (although not necessarily in the bubbly liquid) of a foam column of small or moderate diameter are indeed approximately one-dimensional in the flow regimes employed by foam fractionation. The inherent advantages of one-dimensional descriptions are that they are independent of the cross-sectional area of the column, and therefore its overall throughput. Thus, all the calculations related to sizing the diameter of a foam fractionation column can be done using the one-dimensional theories described in Chapter 3 and then the actual column diameter is determined by the required throughput. That being said, it is unfortunate that this approach has not seen much adoption within the foam fractionation community, which makes the comparison of different-sized columns used by various researchers for various systems rather difficult. Additionally, even in the studies dedicated to column dimensions (Wood and Tran, 1966; Kown, 1971), the total volumetric gas and liquid flowrates were used as the control parameters. The effects observed and attributed to column dimensions were essentially the effects of different superficial gas or liquid velocities, rather than genuine effects caused by scaling of the columns, such as changes in flow regimes or deviations from the one-dimensional flow, which do exist and are discussed below.

We first turn our attention to very small diameter columns, and specifically the lower limit to which the one-dimensional description is valid. Brannigan and Bonfim (2001) conducted drainage experiments in columns of small diameter and concluded that wall effects were insignificant with regard to drainage in columns with an inner diameter of greater than 37.5 mm. Thus, it can be concluded that a one-dimensional description of a pneumatic foam is valid for a column with a diameter greater than this value. That is not to say, however, that foam fractionation cannot be effective in smaller-diameter columns. Indeed, depending on the contact angle of the internal column wall, wall effects can promote net drainage from the foam and therefore enhance the enrichment factor. Thus, there is no minimum limit to the diameter of a foam fractionation column beyond the diameter of the delivered bubbles. However, from a practical point of view, using a cluster of small-diameter columns results in higher capital and operational costs (due to higher frictional pressure drop at the column wall). In some specific cases, such as the production of nisin (see Chapter 10), the increased ratio of the column wall area to liquid volume can cause a greater loss of nisin activity due to adsorption to the column walls. On the other hand, using multiple smaller columns, the risk of column failure is less and it is easier to build in redundancies. In the same example of nisin production, sixteen smaller columns with 50 cm internal diameter were initially built. If one of them fails, for instance, due to leakage of liquid or gas, or blockage of a sparger, the liquid can be evenly distributed into other columns and the process continues while the problematical column is being returned to service.

Turning to the upper limit of the internal diameter of a foam fractionation column, we must consider limitations of foamate collection as well as the possible establishment of convection currents within the device. In Section 7.5, two dominant methods of foamate collection were introduced: The inverted "U" and the launder, the latter being almost universally adopted in protein skimmers and the collection of the concentrate stream in flotation. The foam fractionation unit with the largest known diameter is that described in

Chapter 10 for the production of nisin at over 1.5 m internal diameter, and, in this case, the foam is collected by a contraction at the top of the column leading to a flexible hose in an inverted "U." Thus, a design constraint on the extent of contraction leading to a flexible hose (and therefore the diameter of the column) in this respect is the pressure loss in the flexible hose itself. Another constraint is that of foam stability: In order for a foamate to manifest, the foam must exhibit sufficient stability to reach the top of the column, and it is common experience that foams in smaller-diameter columns can exhibit greater stability than in larger-diameter columns due to the supporting effect of the column wall. Of course, the mechanical strength of the column material and the supporting frame must also be considered, but that is beyond the scope of this book.

If a launder is employed, then this places a genuine constraint on the maximum column cross-sectional area/internal diameter. The foam flows from the top of the tube (sometimes known as the "lip") across a perimeter that goes as a linear function of the column diameter D but the foamate volume that must cross is linear in the cross-sectional area (i.e., goes as D^2). There is therefore an upper limit on the column diameter before foam-handling problems are caused at the top of the unit. This problem is well known in flotation, where froth launders are routinely employed but the column cross-sectional areas tend to be very high, and flotation directs us to a method of solving the problem in the "donut launder" and the "ring launder," both of which serve the purpose of approximately doubling the lip perimeter. Both are devices that crowd the froth outward and inward over a double lip so as to improve froth transport to the launder. Note that a launder can be used only if the foam is wet and flows under gravity, or collapses quickly so the foamate flows, such as in protein skimmers.

Experience from running large-scale foam columns (i.e., approximately greater than 1 m in diameter) indicates that the bottom of the foam layer has shown significant large-scale convective structures, as well as smaller-scale eddies. Convective instabilities in draining foams have been observed in several studies, including that of Hutzler et al. (1998). It is clear that a one-dimensional description is poor at the bottom of the large-diameter column, especially when the foam is wet. However, further up the froth layer as the foam becomes drier as capillary effects relax away and there is a possible change in bubble size distribution, one-dimensional flow is evident. The bottom layer exhibits characteristics of a "well-mixed system," whereas once one-dimensional flow is recovered, the system displays "plug flow" behavior. The root cause of these observations lies in the rheology of the foam: In large-diameter columns, the yield stress is readily exceeded, where the foam is wetter at the bottom of the column, and this results in a departure from one-dimensional plug flow.

In the context of foam fractionation, if a three-dimensional foam flow is established at the bottom of the foam layer because the column is of large diameter, then this is better considered as an extension of the well-mixed bubbly liquid than as part of the rising foam. Thus, the use of a large-diameter foam column has the practical effect of extending the *effective* height of the bubbly liquid, and diminishing the *effective* height of the foam layer. To militate against a well-mixed foam layer that might present, either the column diameter should be decreased or column internals should be judiciously designed to promote one-dimensional flow.

It should be noted that, throughout this book, it has been assumed that the diameter of the column which contains the foam layer is the same as that in which the bubbly liquid resides, although there is no design reason for this beyond ease of column fabrication.

8.2.2 Height

In Chapter 3 the misapprehension that increased height improved enrichment because it engendered greater residence time and therefore more time for liquid to drain from the foam was described. Once capillary effects have dissipated at the very bottom of the column, the only mechanism by which the volumetric liquid fraction can change (unless liquid is introduced into the column) is by a change in bubble size distribution via the mechanisms of film rupture or Ostwald ripening (see Chapter 4 for information on these mechanisms of instability). Clearly, this is a crucially important concept when designing the height of the foam layer in foam fractionation. The question remains as to the effect of increasing the foam layer in foam fractionation.

8.2.2.1 Continuous Foam Fractionation in Simple Enriching Mode with No Change in Bubble Size Distribution

Consider first a column operating continuously in simple enriching mode (i.e., with no external reflux). If there is no change in bubble size distribution with height in the column (so that there is no internal reflux), then there can be only one equilibrium formed between the bubble surface and interstitial liquid in the device, and there can be no change in liquid flux and no change of surface flux with height. Thus, if the bubbly liquid is deep enough to enable adsorption equilibrium to be attained, then under these circumstances, the only purpose that the foam layer serves is to regulate the flux of liquid out of the column. In the circumstance that adsorption equilibrium is not approached in the bubbly liquid, then additional foam height will give additional interface residence time in the unit, and therefore it will take longer for the target species to adsorb. As described in Section 2.7, it is likely that the adsorption rate increases with advection of liquid past the bubble surface, and so the rate of adsorption in the foam layer will be lower than in the bubbly liquid layer, although there are no methods to estimate the rate. A conservative estimate would be to assume that liquid in the foam is effectively stagnant, and adopt the methods described in Chapter 2. That being said, if there is no change in bubble size and no change in liquid flux in the foam layer, then adsorption that occurs in the foam layer merely transports the surface-active material from the interstitial liquid to the bubble surface, both of which eventually report to the product stream, which means the enrichment or recovery is the same with or without the additional adsorption. Thus, it is axiomatic to state that, although the deeper the foam layer, the longer the residence time for adsorption, this is only of consequence if external reflux is employed or if internal reflux occurs autogenously, as is discussed below.

8.2.2.2 Continuous Foam Fractionation in Simple Enriching Mode with a Change in Bubble Size Distribution with Height

The mechanisms by which bubble size distribution can change with height are internal film rupture and Ostwald ripening. If these mechanisms are significant, then the bubble size increases with height and the consequences are twofold:

1. The flux of gas–liquid surface decreases with height in the column, which causes the recovery rate of the adsorbed species to diminish.
2. The flux of interstitial liquid decreases, which causes an increase in enrichment because less of the relatively lean liquid reports to the foamate, and internal reflux (see Section 6.7) is engendered.

It is seen that, in a continuous simple enrichment mode with a change of bubble size distribution as a function of height, increasing the height of the foam layer increases enrichment but decreases recovery. This is an observation that has been made by many researchers in foam fractionation, although sometimes the reason attributed to the observation (i.e., the longer drainage time with increased foam height) is incorrect, as is described in Section 3.1. However, it is stressed that there is no a priori method of determining the change in bubble size as a function of height in foam fractionation. So instead of trying to predict what the optimal foam height is from first principles, bench-scale experiments can be performed to find the optimal gas residence time in foam that gives the desired combination of enrichment and recovery. This optimal gas residence time can be used as a scaling-up parameter.

The velocity of bubbles within the foam layer, V_b, is

$$V_b = \frac{j_g}{1 - \varepsilon} \tag{8.1}$$

Recognize that, because the very bottom of the foam is wetter due to capillarity, the bubbles decelerate over the first several centimeters in the foam, although this effect is small and does not need further consideration by the designer. If the height of the foam layer is H_f, then the residence time within the foam layer, τ_f, is

$$\tau_f = \frac{(1-\varepsilon)H_f}{j_g} \tag{8.2}$$

Note that if the bubble size changes, the liquid fraction will also change and an average value should be used with Equation 8.2. In this case, Equation 8.2 resembles Bikerman's unit of foamability (Equation 4.11), but is different in that Equation 8.2 has applicability to continuously flowing foam, while Bikerman's approach only applies to non-overflowing foam.

8.2.2.3 Continuous Foam Fractionation with Application of External Reflux

As alluded to above, if reflux is applied, the height of the foam layer becomes a material factor in its performance, even if the bubble size distribution remains constant up the column. Thus, under conditions of both external reflux (where enriched liquid is purposefully added to the top of the column) or internal reflux (where reflux is engendered by a change in the bubble size distribution), the enrichment of the foam fractionation device improves with increasing height of the foam layer, so that more theoretical units are manifest. However, there is no better guidance than that of Lemlich (1972), repeated from Chapter 6, that, "Under reasonably favourable operating conditions, an HTU (height of a theoretical unit) on the order of several inches or better can be expected, at least with simple surfactants."

Given the uncertainty in estimating how bubble size changes in a vertical foam column and the height of a theoretical unit, what are required are design heuristics based on experience of operating foam columns, and thus the following advice is offered:

1. In a system with no external reflux, but with a very stable foam layer such that no significant change of bubble size distribution with height is required, a foam layer of no more than approximately 20 cm (although this depends on the gas rate as described by Stevenson, 2007) is required that can fix the liquid flux through the device and to provide a safety margin to ensure that the interface between the bubbly liquid and the foam layer does not approach the top of the column.

2. The recommended height of a system that experiences significant bubble size change with height of the column depends on the process goal. In Section 7.1 it was demonstrated that the performance of a continuous foam fractionator operating in simple enriching mode was highly dependent on bubble size. In general, bigger bubbles mean higher enrichment but lower recovery. The effect of internal reflux, caused by changes in bubble size distribution, acts to further

enhance enrichment. In order to take advantage of the internal reflux, a total height of at least 60 cm is recommended for the reason given above. Any further height will enhance enrichment but reduce recovery in a manner that cannot be predicted from first principles. If the foam stability is too low, such as in most protein skimmers, a stable foam height of 60 cm may not be achievable. In those circumstances, the diameter of the foam column may need to be reduced to increase the superficial gas velocity, and thus final foam height.

3. A system undergoing external reflux but with no significant change in bubble size distribution requires enough foam height to manifest, say, four theoretical units. By adopting a conservative interpretation of Lemlich's advice on the height of a theoretical unit, it is assumed that the HTU is 15 cm (approximately 6 inches) and therefore a total height of approximately 60 cm is required to appropriately benefit from the applied external reflux. It should be noted that the rate of adsorption strongly depends on the properties of the actual system; bench-scale trials may be required to find the exact values of HTU.

In the above discussion, the foam layer was mainly considered. In some cases, such as in simple enriching mode without significant internal bubble coalescence or any external reflux, the enrichment and recovery of the process is determined by the adsorption in the bubbly liquid only, but not the adsorption in the foam layer. Thus, the bubble residence time in the liquid pool is a determining factor. As the bubbly liquid layer is identical to widely used bubble column reactors, the same design principles can be applied. What makes things easier for foam fractionation is that the superficial gas velocity is typically very low and the flow in the bubbly liquid normally resides in the ideal bubbly flow regime, where the slip velocity between the gas bubbles and the liquid is approximately the same as the terminal bubble velocity, that is, $n \approx 1$ in Equation 5.25. This means that the bubble residence time in the bubbly liquid, and thus the adsorption time, is independent of liquid fraction and is a function of column height and the bubble size only.

However, because it is unclear what the adsorption rate enhancement is due to liquid advection past the bubble surface (see Section 2.7), it is difficult to make a priori estimations of adsorption rate, and therefore the required bubbly liquid height, in a foam fractionation column. For industrial design such information must be inferred from smaller test columns where the surface excess of the object substance as a function of adsorption time, thus the minimum time required for adsorption equilibrium, are obtained. When scaling-up a process for industrial use, the adsorption time should be kept at least the same as the value obtained from tests in the bench-top-scale column. If the bubble size used in the large column is bigger, the liquid depth must increase accordingly to maintain a sufficient residence time of the bubbles in the liquid. While a direct measurement of interfacial adsorption at a dynamic interface is not easy, a method based on mass balance is available. Briefly, this method assumes that the interstitial liquid in the foam has the same concentration as the bubbly liquid; by measuring the foamate concentration and the bubble size, the surface excess can be estimated (Wood and Tran, 1966; Lemlich, 1968; Bhattacharjee et al., 1997). Recently, Li et al. (2011a) used such a method with an SDS solution and inferred the mass flux of adsorbed species at the top of the bubbly liquid as a function of bubbly liquid depth. They found that there was no further adsorption of SDS beyond the minimum liquid pool depth of 20 cm that was tested (see Figure 8.1). The mean bubble diameter was 0.415 mm and so, by Clift et al. (2005), the bubble velocity is approximately 10 mm.s^{-1}, and therefore a residence time greater than approximately 20 s has no further advantage. It should be noted that this figure has no general utility because it depends on the adsorption rate of the particular species. For nonionic species, the adsorption extent approximately scales as the square root of the coefficient of molecular diffusion (see Equation 2.17) in a stagnant liquid, which in turn is approximately inversely proportional to the molecular weight via the Stokes-Einstein equation. It will be assumed that a similar dependency holds for a bubble rising through a liquid pool. Thus, the required residence time, and therefore the required bubbly liquid depth, for the approach to adsorptive equilibrium is a weak monotonically increasing function of the molecular weight.

Note that in the method of determining the liquid pool depth described above, the bubbly liquid is assumed to be well mixed with respect to the concentration of surface-active material. This is valid because of the long liquid residence time in the column so the surface-active material has opportunity

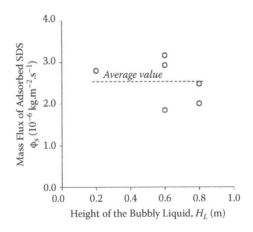

FIGURE 8.1
Adsorbed flux of SDS as a function of height in a bubbly liquid pool. (From Li, X., Evans, G.M., and Stevenson, P. 2011a. *Chemical Engineering Research & Design*, 89, 2298–2308.)

to reach a homogeneous state through diffusion and mixing. In terms of hydrodynamics, the rise of the bubbles can be assumed to be one-dimensional as the bubble residence time is much shorter. This approach is valid for columns of less than 1 meter in diameter. For larger columns, large-scale internal circulation of the liquid may occur, wherein the center of the column the liquid flows co-currently upward with the bubbles, thereby reducing the effective bubble residence time. In this scenario, the design rules for large-scale bubble columns should be applied. Because the bubble size is determined by the sparger design, the sparger and the column height must be designed simultaneously to give the desired adsorption time. There is a large amount of literature readily available (Thorat et al., 1998; Veera and Joshi, 1999; Dhotre et al., 2004), thus it is not repeated herein. If desired, a counter-current flow arrangement (liquid down, gas up) can be used to increase the bubble residence time (Lekang, 2013). Such arrangements are useful if the adsorption is slow but the column height is limited or the bubble size is large.

In addition to the mass transfer and hydrodynamic considerations, in semi-batch foam fractionation, the bubbly liquid layer also serves as a storage vessel for the liquid feed, and therefore it must be deep enough to hold the feedstock if the diameter of the column is limited by other factors. For continuous operations, a minimum liquid retention time in the column is required to reach a certain degree of recovery. The liquid volume in the column must be enough to give the retention time at the desired feed flowrate. These relationships are not specific to foam fractionation and are readily understood by chemical engineers, and therefore will not be elucidated herein.

8.3 Process Control

8.3.1 Operation Control

Chapter 3 stressed that the two primary determinants of the hydrodynamic state of a pneumatic foam were the bubble size distribution and the gas superficial velocity. The bubble size distribution depends on the design of the bubble generation device, the gas rate, the liquid properties (and to a lesser extent gas properties, although the gas employed is almost always air). Thus the only manipulable process variables to control the hydrodynamic condition of the pneumatic foam are (1) the gas superficial velocity, and (2) the bubble size distribution, which itself can only be manipulated by variation of the gas rate if a simple sparger system is employed. Thus, the manipulated process input must be the gas superficial velocity, there being few other genuine options other than control of the liquid feed flowrate in a continuous device as discussed in Section 6.4. However, selection of the controlled variable to maintain at a particular set point is a design decision and may be one of the following:

1. *The liquid fraction in the foam layer of the column, as determined by a differential manometer or a pair of pressure transducers.* The liquid fraction ε is inferred by the differential pressure ΔP measured across a differential height H_p; thus,

$$\varepsilon = \frac{\Delta P}{\rho_f g \Delta H_p} \tag{8.3}$$

where the weight of the gas phase is discounted, ρ_f is the liquid density, and g is the acceleration due to gravity. It must be noted that for a flowing dry foam, the pressure drop caused by the wall shear stress can be significant, and Equation 8.3 must be corrected with a predetermined frictional pressure drop term, ΔP_f, i.e.,

$$\varepsilon = \frac{\Delta P - \Delta P_f}{\rho_f g \Delta H_p} \tag{8.4}$$

ΔP_f can be measured by observing the step change in pressure gradient when a foam suddenly stops rising in a column. For details, please see the work of Li et al. (2011b). Using Equation 8.3 or 8.4, the liquid fraction can be maintained at the set point by controlling the differential pressure using a PID (proportional–integral–derivative) feedback controller manipulating the gas rate with a control valve. There are two important practical considerations: First, the signal from pressure transducers attached to a foam column can exhibit a relatively low signal-to-noise ratio, as was demonstrated in the work of measurements by Shaw et al. (2010). This can either be overcome by extending the integral action of the controller, or by mechanical means: Li et al. (2011b) demonstrated in a laboratory set-up that the signal noise measured from a pair of pressure transducers was largely due to the action of a peristaltic pump feeding a column, and that high-frequency components of the noise could be practically eliminated by placing a buffer tank between the pump and the foam fractionation column. Second, the time to reach steady state in a foam column is long (i.e., transients are long-lived; Shaw et al., 2011) and therefore the controller must be designed with recognition of this fact.

2. *The rate of foamate flow from the top of the column.* In a relatively small semi-batch foam fractionation device, the rate of foamate is straightforward to measure by placing the collection vessel on a balance and taking the rate of change of mass. Then the controlled variable could

be flowrate and the manipulated variable, again, the gas superficial velocity. The time lag between flowrate in the column and its manifestation in the collection vessel could be appreciable, however. In continuous systems, it is inconvenient to measure the mass evolution of collected foamate, so direct measurement of the liquid rate in the foamate would be required; but because the foamate is a two-phase mixture flowing at a low flowrate, this would be challenging indeed to implement. On an industrial scale, if the foam is immediately collapsed to a given foamate, the amount of foamate in a vessel of known cross-sectional diameter can be readily obtained by any conventional level measurement method.

3. The final possible controlled variable is the concentration of the foamate itself. For example, protein concentrations can be measured by UV adsorption, and ionic surfactants can be measured by conductivity, such as the method employed by Stevenson et al. (2011).

8.3.2 Inventory Control

If a device is being operated continuously, active control of the level of the bubbly liquid-foam interface must be adopted. For some foam fractionation applications, such as the recovery of a high added-value product from a fermentation broth, automated level control systems are commercially available at reasonable cost. For example, a pressure transducer coupled with a variable-speed pump or an automated flow control valve can sufficiently keep the liquid level in the column constant. However, in some other applications, such as protein skimming or wastewater treatment, such sophisticated level control systems are not always desirable. Instead, some simple hydrodynamic methods are used. For example, most protein skimmers are inserted directly into the pond; thus the submerged depth determines the hydrostatic pressure at the bottom of the column, and thus the liquid level in the column.

Another method utilizes a device known as the "vented underflow," which maintains the pressure of the liquid pool by raising the outlet level of the bottom rejectate (or underflow) stream, and a schematic representation of such an arrangement is shown in Figure 8.2. The underflow is drawn from the bottom of the column and rises to a T-piece that is a height of H_u above the side draw. One leg of the T is a vertical vent to atmosphere, whereas the other leg is approximately horizontal, which supports the underflow stream. Thus, by manipulating the height of the vented underflow, the gauge pressure at the bottom of the column can be adjusted. This method is particularly convenient if a launder collection vessel is employed; thus the top of the free surface of the foam is at atmospheric pressure. However, if the foam is collected by contracting flow into a much smaller conduit, and particularly if the foam is dry and the conduit is long, the pressure losses at the contraction and the wall of

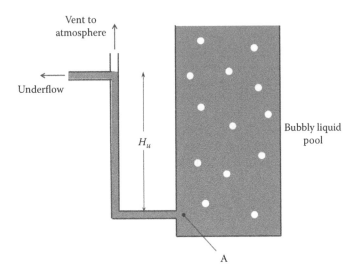

FIGURE 8.2
"Schematic of a vented" underflow arrangement.

the conduit can be appreciable and must be taken into account. In addition, it should be noted that the vented underflow arrangement requires that the surface of liquid in the feed vessel must be below the vented underflow height, meaning that feed cannot be supplied by gravity alone.

8.4 Cost Determinants of Design

Process design is an art, and therefore any process design guidance at all must be subjective and nebulous. This is even truer for foam fractionation than for most other unit operations, as there are so few commercial foam fractionation plants that have been designed. Laboratory units that have been designed tend to be done to investigate a hypothesis rather than for economic coherence. Having said that, most laboratory columns have almost no design at all, given that their purpose tends to be to show whether one species or another is susceptible to foam fractionation (see comment in Section 1.4).

Even the world's largest known commercial foam fractionation installation, described in Chapter 10, came into being more through evolution than design, as was inevitable given the nature of the process and the immaturity of the technology. It is reiterated that full, a priori design is precluded, primarily because of the uncertainties in bubble size distribution and adsorption isotherms in flowing systems.

However, some general guidance can be given based on particular economic features of the process:

1. Extraordinarily low operating cost. Other than the cost of operating feed pumps (and a recycle pump should external reflux be employed) and the cost of gas compression (if conventional spargers are employed), there is no energy consumption. Thus, the cost of energy barely needs to be considered when designing the unit operation.

2. In semi-batch mode, which is favored for most biological processes for reasons of the mode of upstream processes, overall efficiency, and hygiene, the equipment must be cleaned between batches, and this must be taken into account in design. In addition, the necessity of cleaning can introduce a significant labor cost.

3. The position in a typical flowsheet. When operating in enrichment mode, foam fractionation is almost never positioned at the end of a process flowsheet, and is usually used as a cheap method to engender significant enrichment before a purification stage. Thus, there is rarely a process constraint determined by the final product specification.

4. Capital costs are low. Columns are almost always built at ambient temperature and low gage pressure, and mechanical design is straightforward.

5. Foam fractionation is usually just one unit operation in a wider process flowsheet that might include chromatography, spray drying, and centrifugation, all of which can have much larger operating and capital costs. Having said that, downstream processes typically have a low volumetric throughput, which reduces costs, and the ability of foam fractionation to reduce the volumetric duty of downstream processes is one of its primary advantages.

Thus, it is seen that foam fractionation is cheap to build and run in absolute terms, and is even cheaper relative to typical complete flowsheets. Thus, the above comments about cleaning utility aside, there are very few genuine cost determinants of process installation and commissioning.

However, this section is titled "Cost Determinants of Design," and although we have presented the best design methods currently known, they are very bare compared to, say, column distillation. The reason that the nisin plant evolved rather, was designed, was because of the uncertainties described above, and this evolution process can be considered a cost of design. For a device that can approach optimal plant performance, even armed with the methods expounded herein, it is likely that pilot development work would be necessary. Therefore, until heuristic methods for plant-scale design emerge, the most significant costs of foam fractionation can be considered those that are necessary for its development for a specific application.

Nomenclature

Roman

g	Acceleration due to gravity	[m.s^{-2}]
H_f	Height of the foam layer	[m]
ΔH_p	Height difference over which is measured pressure differential	[m]
H_U	Height of the vented underflow	[m]
j_g	Superficial gas velocity within the column	[m.s^{-1}]
ΔP	Pressure differential	[Pa]
P_A	Gauge pressure at the vented underflow draw	[Pa]
V_b	Bubble velocity	[m.s^{-1}]

Greek

ε	Volumetric liquid fraction	[–]
ρ_f	Liquid density	[kg.m^{-3}]
τ_f	Residence time of bubbles in the foam layer	[s]

References

Bhattacharjee, S., Kumar, R., and Gandhi, K.S. 1997. Prediction of separation factor in foam separation of proteins, *Chemical Engineering Science*, 52, 4625–4636.

Brannigan, G., and de Alcantara Bonfim, O.F. 2001. Boundary effects on forced drainage through aqueous foam, *Philosophical Magazine Letters*, 81, 197–201.

Brown, A.K., Kaul, A., and Varley, J. 1999. Continuous foaming for protein recovery. Part I. Recovery of β-casein, *Biotechnology & Bioengineering*, 62, 278–290.

Burghoff, B. 2012. Foam fractionation applications, *Journal of Biotechnology* 161, 126–137.

Clift, R., Grace, J.R., and Weber, M.E. 2005. *Bubble, Drops and Particles*, Dover, ISBN: 0486445801.

Dhotre, M.T., Ekambara, K., and Joshi, J.B. 2004. CFD simulation of sparger design and height to diameter ratio on gas hold-up profiles in bubble column reactors, *Experimental Thermal and Fluid Science*, 28, 407–421.

Hutzler, S., Weaire, D., and Crawford, R. 1998. Convective instability in foam drainage, *Europhysics Letters*, 41, 461–465.

Kown, B.T., 1971. The effects of column height and diameter on the effectiveness of a continuous bubble fractionation system, *Water Research*, 5, 93–102.

Lekang, O.-I. 2013. Protein Skimming, Flotation, Coagulation and Flocculation, in *Aquaculture Engineering*, John Wiley & Sons, New York.

Lemlich, R. 1968. Adsorptive bubble separation methods: Foam fractionation and allied techniques, *Industrial and Engineering Chemistry*, 60, 16–29.

Lemlich, R. 1972. *Adsorptive Bubble Separation Techniques,* Academic Press, Library of Congress Number 75-154398.

Li, X., Evans, G.M., and Stevenson, P. 2011a. Process intensification of foam fractionation by successive contraction and expansion, *Chemical Engineering Research & Design,* 89, 2298–2308.

Li, X., Wang, X., Evans, G.M., and Stevenson, P. 2011b. Foam flowing vertically upwards in pipes through expansions and contractions, *International Journal of Multiphase Flow,* 37, 802–811.

Lockwood, C.E., Bummer, P.M., and Jay, M. 1997. Purification of proteins using foam fractionation, *Pharmaceutical Research,* 14, 1511–1515.

Morgan, G., and Wiesmann, U. 2001 Single and multistage foam fractionation of rinse water with alkyl ethoxylate surfactants, *Separation Science & Technology,* 36, 2247–2263.

Schügerl, K., 2000. Recovery of proteins and microorganisms from cultivation media by foam flotation, in *Advances in Biochemical Engineering/Biotechnology,* Springer-Verlag, Berlin/Heidelberg.

Shaw, R., Evans, G.M., and Stevenson, P. 2011. Start-up transients in a pneumatic foam, *Asia-Pacific Journal of Chemical Engineering,* 6, 613–623.

Stevenson, P. 2007. The hydrodynamic theory of rising foam, *Minerals Engineering,* 20, 282–289.

Stevenson, P., Shaw, R., Tulloch, J., and Evans, G.M. 2011. The influence of stirring upon adsorption of a cationic surfactant (CTAB) to activated carbon particles, *Adsorption Science & Technology,* 29, 99–115.

Stevenson, P., Sullivan, S.P., and Jameson, G.J. 2003. Short-time tracer dispersion in a two-dimensional rising froth, *Chemical Engineering Science,* 58, 5025–5043.

Thorat, B.N., Shevade, A.V., Bhilegaonkar, K.N., Aglawe, R.H., Parasu Veera, U., Thakre, S.S., Pandit, A.B., Sawant, S.B., and Joshi, J.B. 1998. Effect of sparger design and height to diameter ratio on fractional gas hold-up in bubble columns, *Chemical Engineering Research and Design,* 76, 823–834.

Timmons, M.B., Chen, S., and Weeks, N.C. 1995. Mathematical model of a foam fractionator used in aquaculture, *Journal of the World Aquaculture Society,* 26, 225–233.

Veera, U.P., and Joshi, J.B. 1999. Measurement of gas hold-up profiles by gamma ray tomography: Effect of sparger design and height of dispersion in bubble columns, *Chemical Engineering Research and Design,* 77, 303–317.

Wood, R.K., and Tran, T. 1966. Surface adsorption and the effect of column diameter in the continuous foam separation process, *The Canadian Journal of Chemical Engineering,* 44, 322–326.

Wu, C. 2004. *Aquarium Protein Skimmer,* United States Patent US6808625.

9

Process Intensification

9.1 Motivation for Process Intensification in Foam Fractionation

It is a common experience of foam fractionation researchers that, in a conventional foam fractionation column, the various operating parameters have competing effects on the enrichment and recovery of a foam fractionation process (Wood and Tran, 1966; Lemlich, 1968; Timmons et al., 1995). For instance, among the two principal process control parameters (i.e., the superficial gas velocity and the bubble size), an increase in superficial gas velocity increases the flux of surface area and thus improves recovery, but at the same time it results in a higher liquid content of the foam, which reduces the enrichment. Using a smaller bubble size has similar effects on increasing the surface area flux, but also results in a wetter foam that has lower enrichment. While this type of dilemma also exists in other unit operations, for foam fractionation it is particularly complicated due to some intractable properties of the foam, and, to some extent, the lack of a mechanistic understanding of the process.

Beyond the above-mentioned competing effects of operating parameters on enrichment and recovery, there are also intrinsic disadvantages associated with different modes of operation, as briefly mentioned in Chapter 8. For example, Figure 9.1 shows a comparison between the foamate concentration profiles of a batch foam fractionation and that of a continuous process, where the rejectate concentrations of both processes are the same. The area below each curve represents the total amount of objective substance recovered. It can be seen that, because the semi-batch operation has a higher liquid concentration to start with, the average adsorption driving force is higher than in a continuous mode of operation. Consequently, the continuous operation takes more than three times longer to recover the same amount of objective species. However, in some scenarios, such as in wastewater treatment or protein skimming, continuous operation is desirable.

The above constraints and limitations are well understood by most foam fractionation researchers, and choices of appropriate operation parameters are made to have the optimal combination of enrichment and recovery for

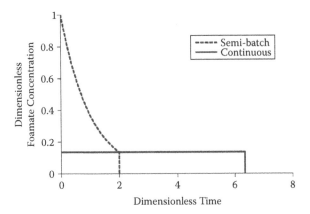

FIGURE 9.1
Foamate concentration profiles of semi-batch foam and continuous fractionation processes.

a given system (Wood and Tran, 1966; Ahmad, 1975; Tharapiwattananon et al., 1996). However, instead of simply making compromises, various process intensification methods have previously been proposed. Most of these methods are based on the fact that the enrichment of surfactant concentration in foam fractionation is a combined effect of interfacial adsorption and foam drainage. Therefore, in order to obtain a higher enrichment, one needs to generate and maintain a large amount of gas–liquid interface while reducing as much of the interstitial liquid as possible. The various process intensification approaches can be classified into two categories based on whether it is meant to enhance interfacial adsorption (adsorption enhancement methods) or to improve interstitial liquid drainage (drainage enhancement methods). The mechanisms and some examples of the implementation of these concepts are discussed in this chapter.

9.2 Adsorption Enhancement Methods

As described in Section 2.2, the relationship between the bulk concentration, C_b, and the equilibrium surface excess, Γ^*, is called an adsorption isotherm and can be generally expressed as $\Gamma^* = f(C_b)$. An example of such a relationship is the Langmuir adsorption isotherm:

$$\Gamma^* = \Gamma_\infty \left(\frac{K_L C_b}{1 + K_L C_b} \right) \tag{9.1}$$

where Γ^* (mol.m^{-2}) is the equilibrium surface excess for a given bulk concentration C_b (mol.m^{-3}); Γ_∞ (mol.m^{-2}) is the saturation surface excess, i.e., the highest possible surface excess when the whole surface is covered by the surface-active molecules; and K_L (m^3.mol^{-1}) is the Langmuir constant. By increasing C_b, the equilibrium surface excess can be increased until $\Gamma^* \to \Gamma_\infty$, as long as the exposure time is long enough to attain an approach to equilibrium. Generally speaking, the surface-active species concentration in the feedstock of a foam fractionation process is very low; thus, the equilibrium surface excess Γ^* is below the saturation value Γ_∞. In this case, process intensification can be achieved by increasing the concentration of the liquid that the gas–liquid interface is in equilibrium with. Depending on where in the column the adsorption of surface-active molecules occurs, the extent of gas–liquid surface adsorption can be enhanced by either (1) multistage or (2) reflux processing. As a general rule, if adsorption occurs mainly in the bubbly liquid, a multistage approach can be effective; while if the adsorption occurs mainly in the foam layer, reflux approaches are adopted.

9.2.1 Multistage Processing

Multistage processing includes foam fractionation devices with physically distinct stages, either in the compact form of a single column with internal components that resemble those found in distillation columns (Criswell, 1976; Leonard and Blacyki, 1978; Schügerl, 2000; Boonyasuwat et al., 2003; Darton et al., 2004; Boonyasuwat et al., 2005, 2009; Rujirawanich et al., 2012) or a series of conventional columns connected with pipes and fittings (Morgan and Wiesmann, 2001; Lekang, 2013). Depending on the foam stability exhibited by a particular system, it may be necessary to employ mechanical foam breakers between stages. Both designs are based on the assumption that adsorption occurs mainly in the liquid pool, and that only one equilibrium stage of adsorption can be achieved in a conventional column. Therefore, a higher surface excess can be achieved by repeated foaming cycles.

A typical compact multistage foam fractionation device is shown in Figure 9.2; it was developed by Darton et al. (2004). The foam is destroyed by mechanical paddles between each stage, and the resultant foamate is mixed with the feed in order to increase the bulk concentration. The horizontal multistage bubble fractionator of Leonard and Blacyki (1978) shown in Figure 9.3 is based on the same principle as the vertical column except that the stages are arranged in a horizontal device. Feed is introduced in an intermediate stage, and liquid pools of adjacent stages are connected by a horizontal tube located near the bottom. Hydrostatic pressure causes the liquid to flow from one stage to the next. The foam flows horizontally in the opposite direction to the liquid. No foam-breaking mechanism is employed so the device only has utility for foam with a low stability, and is specifically nominated as an intensification of the bubble fractionation process. Consequently, the foam layer is shallower compared to conventional foam fractionation columns.

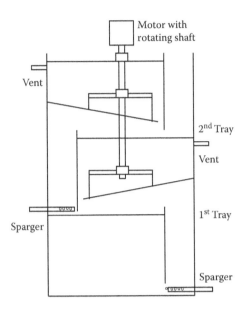

FIGURE 9.2
Compact multistage foam fractionator. (Adapted from Darton, R.C., Supino, S., and Sweeting, K.J. 2004. Development of a multistaged foam fractionation column, *Chemical Engineering and Processing*, 43, 477–482.)

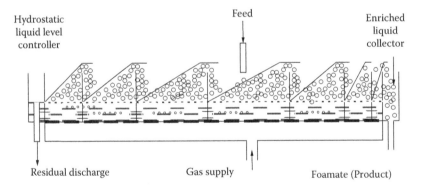

FIGURE 9.3
Horizontal-multistage foam (bubble) fractionation device. (Adapted from Leonard, R.A., and Blacyki, J.D. 1978. *Industrial and Engineering Chemistry Process Design and Development*, 17, 358–361.)

While such compact multistage columns (Leonard and Blacyki, 1978; Schügerl, 2000; Boonyasuwat et al., 2003; Darton et al., 2004; Boonyasuwat et al., 2005, 2009; Rujirawanich et al., 2012) do have their advantages, they typically have complicated internal designs and multiple moving parts. This can put long-term reliability at risk and make routine maintenance difficult.

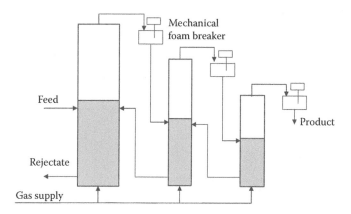

FIGURE 9.4
Multistage foam fractionation by operating individual columns in series.

Thus, instead of using a compact design, exactly the same effect can also be achieved by operating individual columns in series, as shown in Figure 9.4. In this arrangement, the foamate from one stage is used as the feed to the next stage, while the rejectate is returned to a previous stage. An application of such a concept in protein skimming was mentioned by Lekang (2013), albeit with only two stages.

The concept illustrated in Figure 9.4 should not be confused with the multistage devices developed by Morgan and Wiesmann (2001), one of which is shown in Figure 9.5. While all the multistage devices discussed above are specifically designed to enhance interfacial adsorption by increasing the surfactant concentration in the bubbly liquid, the design by Morgan and Wiesmann was intended to overcome the intrinsic disadvantage of continuous operation, which was briefly discussed in Chapter 8 and the beginning of this chapter. The foamate concentration profile of Morgan and Wiesmann's design is compared with that of a single semi-batch operation in Figure 9.6. It can be seen that, by using multiple columns in series, it is possible to approach the performance efficiency of a semi-batch process by operating multiple continuous foam fractionation columns in series. However, to make the best of this design, a detailed material balance and hydrodynamics calculation, coupled with sophisticated column dimensions design, are required, which the original work of Morgan and Wiesmann (2001) unfortunately lacked.

9.2.2 Reflux Processing

Multiple equilibrium stages can also be engendered within a continuous foam layer by either returning some of the foamate to the foam layer as an external reflux stream (de Lucena et al., 1996; Yamagiwa et al., 2001; Stevenson and Jameson, 2007; Martin et al., 2010), or autogenously by bubble coalescence

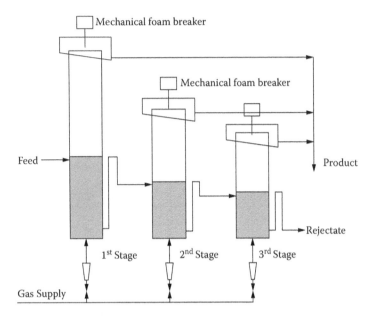

FIGURE 9.5
Multistage-conventional foam fractionators. (Adapted from Morgan, G., and Wiesmann, U. 2001. *Separation Science and Technology*, 36, 2247–2263.)

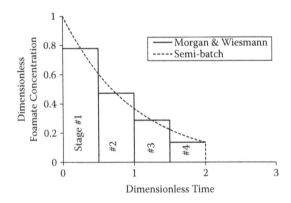

FIGURE 9.6
Foamate concentration profile of a single-column semi-batch operation compared with the multistage design in Figure 9.5.

causing internal reflux (Wood and Tran, 1966; Lemlich, 1968; Stevenson et al., 2008). This was discussed in great detail in Chapter 6, together with dem-onstrations of how the hydrodynamic theory of rising foam proposed by Stevenson (2007) can be employed to describe such processes; thus it is not repeated herein.

As discussed above, both multistage and reflux approaches are based on the same principle that the interfacial adsorption can be enhanced by increasing the surfactant concentration in the liquid that is in contact with bubbles. In terms of device design, both approaches require that the foam be effectively collapsed, and the only difference is whether the foamate is returned to the foam layer or the bubbly liquid. Therefore, there is no reason that the two methods cannot be combined. One way to combine them is by directing the foamate from an upstream stage to the foam layer of the downstream stage, as shown in Figure 9.7. If required, the rejectate from each stage can be collected separately for higher throughput (Criswell, 1976).

As discussed in Chapter 8, for reflux methods to be effective, a certain foam height must be maintained. However, this is not always possible, as some systems may not generate stable foam at all; thus only the multistage approaches can be used. Indeed, where multistage foam fractionation is employed, it is generally because the surfactant concentration in the feedstock is very low. If the surfactant concentration is very high and the surface excess is close to the saturation value, there is no need for either multistage or reflux processing. However, when the foam is indeed stable enough, it is recommended that external reflux is used, although the power consumption for foam breaking can become significantly higher as the foam becomes more and more stable. In the case of the processing of biomaterials, such as proteins and peptides, multiple foam generation/bursting cycles associated with multistage or reflux operations may also cause serious denaturation of the target molecule (Liu et al., 1998; Clarkson et al., 1999a, 1999b).

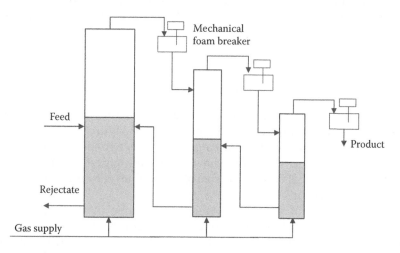

FIGURE 9.7
A combined multistage-external reflux concept.

9.3 Drainage Enhancement Methods

Discussed above were various multistage or reflux methods for improving interfacial adsorption. These methods are better suited for systems with very low surfactant concentrations where the foam is easy to collapse. There is a limitation, however, that when the surfactant concentration is very high, the bubble surface is nearly saturated. In this case, the foam is typically also very stable, which makes multistage or reflux approaches less attractive as both require efficient collapse of the foam. It is noted that anti-foam cannot be used with multistage or reflux approaches, as discussed elsewhere. Thus, the only possible method to further enhance the enrichment is by reducing the liquid content of the foam, that is, by enhanced liquid drainage. Although a dry foam can be obtained using a very slow superficial gas velocity, it may not be economically viable because it results in low throughput.

By recalling the hydrodynamic theory of pneumatic foam expounded upon in Chapter 3, the superficial drainage velocity in the vertical direction, j_d, can be estimated from the following empirical relationship:

$$j_d = \frac{\rho_f g r_b^2}{\mu} m\varepsilon \qquad (9.2)$$

where ρ_f and μ are the density and dynamic viscosity of the interstitial liquid, respectively; g is the acceleration due to gravity; the product of $\rho_f g \varepsilon$ is the pressure gradient in the foam, which is the driving force of foam drainage in the vertical direction; r_b is an average bubble size; ε is the liquid fraction of the foam in question; and m and n are two adjustable parameters that are constants for a given system. It can be seen that the only methods to increase drainage rate without changing liquid properties or increasing the liquid fraction are by increasing either the bubble size or the drainage driving force by some other means.

9.3.1 Drainage Enhancement by Bubble Size Manipulation

Recall that the hydrodynamic state of a vertical pneumatic foam is strongly dependent on the characteristic bubble size (see Figure 3.6), and that the enrichment and recovery from a single foam fractionation column shares the same dependency. Thus, there have been some suggested process intensification methods that control the size of the bubbles entering the foam layer using column internals.

Bando et al. (2000) simply inserted a draft tube in the bubbly liquid pool and sparged gas bubbles through the draft tube (see Figure 9.8), in a similar manner to that of an air-lift reactor. A liquid convection pattern is established up through the draft tube, and around and down the annulus or "downcomer." Thus, smaller bubbles are recirculated to the bottom via

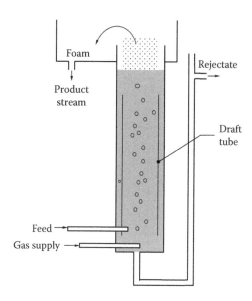

FIGURE 9.8
The draft tube apparatus of Bando et al. (Bando, Y., Kuze, T., and Sugimoto, T. 2000. *Korean Journal of Chemical Engineering*, 17, 597–599.)

the downcomer, and only relatively larger bubbles join the foam layer above. Recall that larger bubbles in the foam later mean a drier foam, a lower flux of interstitial liquid but with a lower surface flux, and therefore higher enrichment and lower recovery rate. They used their device, operating continuously, to recover metal ions from an aqueous wastewater stream (note that this is sometimes referred to as "ion flotation"). They found that the use of the draft tube decreased the liquid flux and recovery rate but increased enrichment.

However, there is nothing surprising about this result when viewed from the perspective of the theory expounded in Chapter 3. The draft tube excludes small bubbles from the foam layer, and thus the characteristic size within the foam becomes larger. Consequently, the liquid flux is lower but the enrichment is higher. Precisely the same hydrodynamic effect could have been achieved by redesigning the sparger to create larger bubbles, although it should be said that the draft tube may engender better mass transfer in the bubbly liquid pool. Note also in Figure 9.8 that Bando utilized a form of a vented underflow (see Section 8.3.2) for liquid inventory control.

The use of perforated plates to reduce liquid flux has also been investigated by Aguayo and Lemlich (1974) and Tsubomizu et al. (2003). A schematic diagram of the foam fractionator used by the latter to enrich a solution of polyvinyl alcohol is shown in Figure 9.9. A series of four perforated plates with circular orifices of diameter 2 mm and perforated area of 3% of the column cross-section were placed within the foam later that was subject to

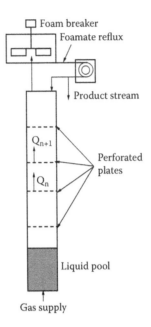

FIGURE 9.9

Foam fractionator with perforated plates in the foam layer as drawn by Tsubomizu et al. (From Tsubomizu, H., Horikoshi, R., Yamagiwa, K., Takahashi K., Yoshida M., and Ohkawa, A. 2003. *Journal of Chemical Engineering Japan*, 36, 1107–1110. With permission.)

external reflux. The liquid fraction within the foam was observed to decrease due to flow through the perforated plates, and enhanced enrichment was the result. The authors postulated that part of the foam lamellae is "partially wiped" by the orifices as the bubbles squeeze through. Such an explanation is reminiscent of the conceptualization of a flotation froth proffered by Finch et al. (2006), who suggest that there is a layer of "bound" liquid around a bubble that moves up the froth, surrounded by "free" liquid that moves downward; this notion was shown by Stevenson and Evans (2008) to be unlikely because the liquid cannot support step changes in strain rate. In fact, the reduced liquid fraction caused by the perforated plates is much more likely to be due to changes in bubble size due to coalescence caused by flow through the orifice; this could have engendered internal reflux if there were a way for the liquid to be released to the foam beneath the plates. However, for a plate with 2 mm diameter orifices and only 3% open area, the liquid is unlikely to be able to flow downward through the orifices, so the liquid should accumulate on the upper surface of the plates. Tsubomizu et al. did not make such an observation but stated that volumetric liquid flux on a lower stage, Q_n, is greater than that on a higher stage, Q_{n+1}, which does not satisfy mass balances. However, the study did prove that there is a simple method to enhance the rate of change of bubble size distribution in a column. If there is a clear path for the interstitial liquid to return to the foam layer, this should promote internal reflux. In the production of nisin, as will be described in Chapter 10, various internals were tested to improve drainage, including metal meshes of various sizes. They were found to marginally improve enrichment at the expense of lower recovery rates (i.e., a longer processing time is required to achieve the same total recovery in a semi-batch operation).

It should be noted that Aguayo and Lemlich's (1974) study found that columns with perforated plates were no better than conventional columns at low superficial gas velocities. It is possible that the bubble coalescence that apparently caused the success of Tsubomizu's unit was governed by inertial effects that did not manifest in the earlier experiments at low flowrates.

In addition to bubble size manipulation using internals, Wang et al. (2010) used a vacuum to enlarge the bubbles within the foam and observed higher enrichment at the expense of lower recovery. Their data showed that exactly the same enrichment and recovery can be obtained by using a lower super-ficial gas velocity instead of using bigger bubbles. By theoretical simulations invoking the theory of Stevenson (2007), Li (2011) showed that, if there is no further interfacial adsorption, any improvement in the enrichment ratio that can be obtained using a larger bubble size can also be obtained using a lower superficial gas velocity. Thus, unless internal reflux is engendered (which again makes the method an adsorption enhancement method), manipula-tion of bubble size will not produce a better performance than by simply employing a lower superficial gas velocity.

9.3.2 Drainage Enhancement by Flow Manipulation

As described above, by simply modifying the bubble size, the flux of inter-stitial liquid is reduced at the expense of losing some desirable surface flux, which has the same effect as using a lower superficial gas velocity; thus, no process intensification is achieved. It is appropriate to say that process inten-sification by drainage enhancement can only be accomplished by reducing the liquid flux *without* compromising surface flux. Both of the current authors independently devised column internals that could do this by altering the flow of the foam.

The first method harnessed the Boycott (1920) Effect for enhanced sedi-mentation down inclined solid surfaces. Boycott observed that blood cor-puscles sedimented out of the liquid in a test tube if the tube was oriented on a slant, because the particles needed to settle only to the wall of the tube whence they could slide down the tube wall to the bottom. Thus, sys-tems that embody inclined plates have enhanced phase segregation utility. This feature is employed in the so-called Reflux Classifier device (Laskovski et al., 2006) for the separation of particle species from the elutriate of a flu-idized bed. The same effect was exploited by several foam fractionation researchers to enhance foam drainage (Criswell, 1976; Zeng et al., 2002; Wang et al., 2013). Criswell (1976) found that the inclined angle affects the performance of the column when used in protein foam fractionation. He observed that, at 45°, the inclined column performed worse than a vertical column because serious internal coalescence occurred but the resultant liquid quickly drained to the wall of the column without much contact with the bubble, which eliminated the benefit of internal reflux. Instead, at 22.5°, better results were obtained. Dickinson et al. (2010) took it a step further to include parallel inclined channels (PICs) (20° to vertical), as shown in the inset of Figure 9.10. The liquid in the foam needs only to drain to the topside of an inclined plate before it runs down the plate as a stream to the bubbly liquid. Experiments to enrich a solution of the ionic surfactant cetyltrimethylammonium bromide (CTAB) showed that values of enrichment were up to four times greater than in a plane vertical column with no internals. However, there are practical problems with the device drawn in Figure 9.10. In the case of recovery of surface-active proteins, the increased column wall surface-to-foam volume ratio in a PIC column could cause greater loss of the product, as a significant amount of the protein molecules can adsorb onto the surface of the plates that is in contact with the foam (Joosten and Nuñez, 1995). Therefore, special treatment of the plate surface may be required if this method is applied to foam fractionation of proteins that strongly adsorb. Additionally, there might be mechanical difficulties in implementing the device drawn in Figure 9.10 at large scale, given that the section containing the inclined plates is orientated at a jaunty angle. However, the same effect can be engendered using baffles in a vertical column as internals (Yang et al., 2011).

While another of the current authors (Li) was at Hebei University of Technology, he developed the "foam riser device" as described by Li et al. (2011a). The simple drainage enhancement device, a photograph of which is

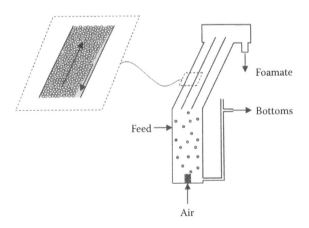

FIGURE 9.10
The inclined plate foam fractionation device of Dickinson et al. (From Dickinson, J.E., Laskovski, D., Stevenson, P., and Galvin, K.P. 2010. *Chemical Engineering Science*, 65, 2481–2490.)

given in Figure 9.11 with a schematic representation in Figure 9.12, consists of a foam riser mounted in the center of a plate. The device creates a contraction of flow area of the foam, followed by an expansion. Above the expansion outlet there is a slip-surface that takes the approximate form of an inverted truncated cone, where the bubbles in the foam travel on paths inclined to

FIGURE 9.11
Photograph of a single foam riser.

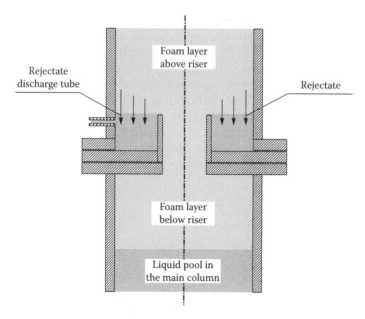

FIGURE 9.12
Schematic representation of the bottom-most foam riser in a column.

various degrees based on their positions in the foam, which is somewhat similar to the inclined plates described above. Below the slip-surface there is a stagnant zone that creates a sink into which liquid drains, and therefore the flux of liquid leaving the plate is reduced, as described in Section 3.8, via the theory of Li et al. (2011b). The rejected liquid forms on the top side of the plate as shown in Figure 9.12, and this can be drained and recycled, or simply returned to the bubbly liquid beneath the foam.

Li et al. (2011a) found that a single foam riser plate device could, in some circumstances, reduce the liquid flux up the column in a relatively wet foam stabilized with SDS by 35% with no discernible reduction in surface flux. A relatively low improvement in enrichment was observed due to a single riser plate in SDS foam because of the nature of the adsorption isotherm. However, for the enrichment of proteins, the utility would be significantly greater, especially if multiple riser plates were employed in series.

Both drainage enhancement devices described above are effective mainly for wet foams where the drainage rate is relatively high. They have a common characteristic in that they enable a shortcut route for the interstitial liquid to drain back to the bubbly liquid. Neither of them have any significant impact on dry foams in which the drainage rate is relatively low.

9.3.3 Drainage Enhancement by Other Methods

Under gravity, the driving force of drainage is $\rho_f g \varepsilon$, which diminishes as the liquid fraction approaches zero. In this case, none of the above-mentioned is particularly effective for further reducing the liquid content, unless an external source of driving force is applied. Khaskova and Kruglyakov (1995) employed a cartridge filter with reduced pressure to draw part of the interstitial liquid out of a dry foam, but such an arrangement is unlikely to have any practical utilization. Centrifugation might be able to further dry a foam, if the centrifugation force is not too high that the foam is collapsed. Reducing the viscosity of the interstitial liquid by increasing the temperature of the foam might be effective (Yan et al., 2011) if the additional power consumption is justified and the high temperature does not cause serious denaturation of the objective substance.

9.4 Concluding Remarks

Studies on the process intensification of foam fractionation are still highly empirical. Although some adsorption enhancement approaches have been modeled and the mechanisms seem clearly understood, these have been based on sets of simplifying assumptions. For example, in the compact multistage designs (Figure 9.2), immediate adsorption equilibrium is assumed at each stage and therefore the dimensions of the column, including the internal structure, are essentially arbitrary numbers. Drainage enhancement appears more promising because it can improve the enrichment without sacrificing the gas–liquid interfacial area, but such approaches rely on drainage so they have little utility for dry foams. Our recommendation is that for very stable dry foam, if space is not limited, using a larger column diameter, at least for the foam layer, might be more effective than any other method with regard to improving enrichment. However for unstable foam that can be easily collapsed, multistage treatment may engender better results. Internal or external reflux should also be considered if the foam is of intermediate stability, i.e., if the foam is stable enough to support a proper foam layer but not too power-consuming to break it. For high-throughput processes of relatively wet foam, one of the drainage enhancement methods can be considered.

Nomenclature

Roman

C_b	Bulk concentration	[mol.m^{-3}]
g	Acceleration due to gravity	[m.s^{-2}]
j_d	Liquid drainage superficial velocity	[m.s^{-1}]
K_L	A constant in the Langmuir isotherm	[m^3.mol^{-1}]
m	Dimensionless number used in Equation 9.2	[–]
n	Dimensionless index used in Equation 9.2	[–]
r_b	Bubble radius	[m]

Greek

ε	Volumetric liquid fraction in the foam	[–]
Γ_∞	Saturation surface excess	[mol.m^{-2}]
Γ^*	Equilibrium surface excess	[mol.m^{-2}]
μ	Interstitial liquid dynamic viscosity	[Pa.s]
ρ_f	Liquid density	[kg.m^{-3}]

References

Aguayo, G.A., and Lemlich, R. 1974. Countercurrent foam fractionation at high rates of throughput by means of perforated plate columns, *Industrial & Engineering Chemistry Process Design & Development*, 13, 153–159.

Ahmad, S.I. 1975. Laws of foam formation and foam fractionation. I. The effect of different operating parameters on the foam fractionation of albumin from a solution containing organic and inorganic materials, *Separation Science*, 10, 673–688.

Bando, Y., Kuze, T., and Sugimoto, T. 2000. Development of a bubble column for foam separation, *Korean Journal of Chemical Engineering*, 17, 597–599.

Boonyasuwat, S., Chavadej, S., Malakul, P., and Scamehorn, J.F. 2003. Anionic and cationic surfactant recovery from water using a multistage foam fractionator, *Chemical Engineering Journal*, 93, 241–252.

Boonyasuwat, S., Chavadej, S., Malakul, P., and Scamehorn, J.F. 2005. Surfactant recovery from water using a multistage foam fractionator. Part I. - Effects of air flow rate, foam height, feed flow rate and number of stages, *Separation Science and Technology*, 40, 1835–1853.

Boonyasuwat, S., Chavadej, S., Malakul, P., and Scamehorn, J.F. 2009. Surfactant recovery from water using a multistage foam fractionator: Effect of surfactant type, *Separation Science and Technology*, 44, 1544–1561.

Boycott, A.E. 1920. Sedimentation of blood corpuscles, *Nature*, 104, 532.

Brown, A.K., Kaul, A., and Varley, J. 1999. Continuous foaming for protein recovery. Part I. Recovery of β-casein, *Biotechnology & Bioengineering*, 62, 278–290.

Clarkson, J.R., Cui, Z.F, and Darton, R.C. 1999a. Protein denaturation in foam. I. Mechanism study, *Journal of Colloid and Interface Science*, 215, 323–332.

Clarkson, J.R., Cui, Z.F., and Darton, R.C. 1999b. Protein denaturation in foam. II. Surface activity and conformational change, *Journal of Colloid and Interface Science*, 215, 333–338.

Criswell, L.G. 1976. Method of Separating and Recovering Soluble Proteins from Protein Containing Solutions Employing Foam Fractionation. United States Patent US3969336.

Darton, R.C., Supino, S., and Sweeting, K.J. 2004. Development of a multistaged foam fractionation column, *Chemical Engineering and Processing*, 43, 477–482.

de Lucena, S.L., Miranda, E.A., and Santana, C.C. 1996. The effect of external reflux on the foam fractionation of proteins, *Applied Biochemistry and Biotechnology*, 57–58, 57–65.

Dickinson, J.E., Laskovski, D., Stevenson, P., and Galvin, K.P. 2010. Enhanced foam drainage using parallel inclined channels in a single-stage foam fractionation column, *Chemical Engineering Science*, 65, 2481–2490.

Finch, J.A., Gelinas, S., and Moyo, P. 2006. Frother-related research at McGill University, *Minerals Engineering*, 19, 726–733.

Galvin, K.P. 2006. Segregation of hydraulically suspended particles in inclined channels, *Chemical Engineering Science*, 61, 7269–7278.

Joosten, H.M.L.J., and Nuñez, M. 1995. Adsorption of nisin and enterocin 4 to polypropylene and glass surfaces and its prevention by Tween 80, *Letters in Applied Microbiology*, 21, 389–392.

Khaskova, T.N., and Kruglyakov, P.M. 1995. Regularities in the concentration and separation of surfactants in a foam, *Russian Chemical Reviews*, 64, 235–247.

Laskovski, D., Duncan, P., Stevenson, P., Zhou, J., and Galvin, K.P. 2006. Segregation of hydraulically suspended particles in inclined channels, *Chemical Engineering Science*, 61:7269–7278.

Lekang, O.-I. 2013. Protein Skimming, Flotation, Coagulation and Flocculation, in Aquaculture Engineering, John Wiley & Sons, New York.

Lemlich, R. 1968. Adsorptive bubble separation methods: Foam fractionation and allied techniques, *Industrial and Engineering Chemistry*, 60, 16–29.

Leonard, R.A. and Blacyki, J.D. 1978. Multistage bubble fractionator. *Industrial and Engineering Chemistry Process Design and Development*, 17, 358–361.

Li, X. 2011. Comments on "Intensified effect of reduced pressure on the foam fractionation process of Bovine Serum Albumin" of Wang et al., *Separation Science and Technology*, 46, 1857–1860.

Li, X., Evans, G.M., and Stevenson, P. 2011a. Process intensification of foam fractionation by successive contraction and expansion, *Chemical Engineering Research & Design*, 89, 2298–2308.

Li, X., Wang, X., Evans, G.M., and Stevenson, P. 2011b. Foam flowing vertically upwards in pipes through expansions and contractions, *International Journal Multiphase Flow*, 37, 802–811.

Liu, Z., Liu, Z., Wang, D., Ding, F., and Yuan, N. 1998. On the denaturation of enzymes in the process of foam fractionation, *Bioseparation*, 7, 167–174.

Martin, P.J., Dutton, H.M., Winterburn, J.B., Baker, S., and Russell, A.B. 2010. Foam fractionation with reflux, *Chemical Engineering Science*, 65, 3825–3835.

Morgan, G., and Wiesmann, U. 2001. Single and multistage foam fractionation of rinse water with alkyl ethoxylate surfactants, *Separation Science and Technology*, 36, 2247–2263.

Rujirawanich, V., Chuyingsakultip, N., Triroj, M., Malakul, P., and Chavadej, S. 2012. Recovery of surfactant from an aqueous solution using continuous multistage foam fractionation: Influence of design parameters, *Chemical Engineering and Processing: Process Intensification*, 52, 41–46.

Schügerl, K. 2000. Recovery of proteins and microorganisms from cultivation media by foam flotation, in *Advances in Biochemical Engineering/Biotechnology*, Springer-Verlag, Berlin/Heidelberg.

Stevenson, P. 2007. The hydrodynamic theory of rising foam, *Minerals Engineering*, 20, 282–289.

Stevenson, P., and Jameson, G.J. 2007. Modelling continuous foam fractionation with reflux, *Chemical Engineering and Processing: Process Intensification*, 46, 1286–1291.

Stevenson, P., and Evans, G.M. 2008. Comments on "Frother-related research at McGill University" by Finch, Gelinas, and Moyo, *Minerals Engineering*, 21, 241–242.

Stevenson, P., Li, X., and Evans, G.M. 2008. A mechanism for internal reflux in foam fractionation, *Biochemical Engineering Journal*, 39, 590–593.

Tharapiwattananon, N., Scamehorn, J.F., Osuwan, S., Harwell, J.H., and Haller, K.J. 1996. Surfactant recovery from water using foam fractionation, *Separation Science and Technology*, 31, 1233–1258.

Timmons, M.B., Chen, S., and Weeks, N.C. 1995. Mathematical model of a foam fractionator used in aquaculture, *Journal of the World Aquaculture Society*, 26, 225–233.

Tsubomizu, H., Horikoshi, R., Yamagiwa, K., Takahashi, K., Yoshida, M., and Ohkawa, A. 2003. Effect of perforated plate on concentration of poly(vinyl alcohol) by foam fractionation with external reflux, *Journal of Chemical Engineering Japan*, 36, 1107–1110.

Wang, J., Liu, G., Wu, Z., and Zhang, L. 2010. Intensified effect of reduced pressure on the foam fractionation process of bovine serum albumin, *Separation Science and Technology*, 45, 2489–2496.

Wang, Y, Wu, Z, Li, R, and Zhang, Z. 2013. Enhancing foam drainage using inclined foam channels of different angles for recovering the protein from whey wastewater, *Colloids and Surfaces A: Physicochemical and Engineering Aspects*, 419, 28–36.

Winterburn, J.B., and Martin, P.J. 2009. Mechanisms of ultrasound foam interactions, *Asia-Pacific Journal of Chemical Engineering*, 4, 184–190.

Wood, R.K., and Tran, T. 1966. Surface adsorption and the effect of column diameter in the continuous foam separation process, *The Canadian Journal of Chemical Engineering*, 44, 322–326.

Yamagiwa, K., Iwabuchi, M., Tsubomizu, H., Yoshida, M., and Ohkawa, A. 2001. Effect of external reflux on foam fractionation of poly(vinyl alcohol), *Journal of Chemical Engineering of Japan*, 34, 126–131.

Yan, J., Wu, Z., Zhao, Y., and Jiang, C. 2011. Separation of tea saponin by two-stage foam fractionation, *Separation and Purification Technology*, 80, 300–305.

Yang, Q.W., Wu, Z.L., Zhao, Y.L., Wang, Y., and Li, R. 2011. Enhancing foam drainage using foam fractionation column with spiral internal for separation of sodium dodecyl sulphate, *Journal of Hazardous Materials*, 192, 1900–1904.

Zeng, W., Li, H., Li, B., Cong, W., Cai, Z., and Ouyang, F. 2002. Harvesting of spirulina cells by foam flotation on a laboratory scale inclined column, *Journal of Chemical Industry and Engineering (Chinese)*, 53, 918–923.

10

Case Study: The Production of Nisin

10.1 Production and Application of Nisin

Nisin is an antimicrobial peptide produced by some strains of *Lactococcus lactis*, which is effective against a wide range of G+ bacteria. It is present in milk and other food due to the natural existence of lactic acid bacteria (LAB); thus, human beings had been exposed to nisin for many centuries before it was identified in the 1920s. Because nisin is a peptide, it is digested and inactivated in the human stomach and has no detrimental effect on human health. In 1969, nisin was approved by the Joint FAO/WHO Expert Committee on Food Additives and was given the food additive number 234. It is currently permitted for use in food products, including dairy products, egg products, meat products, canned foods, etc., in over 50 countries (Delves-Broughton et al., 1996; Lück and Jager, 1997; Delves-Broughton, 2005; EFSA, 2006a). In the formative period of production, nisin was referred to as an "antibiotic" (Hirsch, 1950; Liu and Hansen, 1990) but that was inappropriate. Instead, nisin is a member of a group of polypeptides with the new name of "bacteriocins," which are primary metabolites (antibiotics are secondary metabolites) produced by bacteria that kill or inhibit the growth of other bacteria (Delves-Broughton et al., 1996; Abee and Delves-Broughton, 2003). Unlike antibiotics, bacteriocins, including nisin, do not have any toxicity or side effects.

Nisin is prepared and sold as a dry powder with a white to light-brown color. Its activity, or so-called potency, is measured in International Units (IU). Commercial nisin products contain approximately 1,000 IU per milligram of dry powder, which is 2.5% (w/w) pure nisin. The remainder is salt and other proteins derived from the fermentation broth. Conventionally, the fermentation medium contains milk protein or milk derivatives. However, for applications where milk proteins are prohibited, such as the use of nisin in liquid egg products, alternative fermentation media, such as those that are sugar based, must be used (EFSA, 2006b).

Nisin has excellent stability in its dry form, and commercial products are normally labeled with a shelf-life of 2 years. In solution, the stability of nisin depends on the pH and temperature. The optimum pH for the highest nisin stability is found to be around pH 2.0 to 3.0. Low temperature is preferable for the retention of nisin activity (Liu and Hansen, 1990). When mixed with food components, various compounds in foods can protect nisin molecules to some extent; thus, the pasteurizing temperature (~65°C) is rarely a problem (Abee and Delves-Broughton, 2003). The autoclaving temperature (121°C) could severely inactivate nisin at pH values other than 3 (Davies et al., 1998).

Commercial nisin is produced in a liquid medium by batch fermentation. Improvements to the simple batch process have attracted continued interest from both the industry and in academia. These include various modifications to the fermentation medium and the application of a fed-batch fermentation technique to overcome substrate inhibition (Hurst and Dring, 1968; Bertrand et al., 2001; Penna and Moraes, 2002; Lv et al., 2004a, 2004b; Lv et al., 2005; Wu et al., 2009). To overcome potential product inhibition, online removal of nisin from the fermentation broth was attempted (Pongtharangku and Demirci, 2007; Liu et al., 2010), but industrial-scale application of such techniques has not been reported.

Figure 10.1 shows the nisin titer profile during the entire course of a simple batch-mode fermentation process. The fermentation was carried out in a benchtop 1-L, continuously stirred tank reactor (CSTR) at 30°C. The media consisted of corn syrup and sucrose as the carbon source and peptone as the nitrogen source. The pH was maintained at above 6.25 during the course of the fermentation. Synthesis of nisin started in the early exponential phase of growth (4 to 6 hr) of the bacterium, reached a maximum toward the end of this phase (10 to 12 hr), and ceased when entering the stationary phase (after approximately 13 hr). There was a substantial decrease in nisin titer after the peak, which was due to proteolytic degradation and adsorption of nisin onto the bacteria cells (Hurst and Dring, 1968; Van't Hul and Gibbons, 1996; Parente and Ricciardi, 1999).

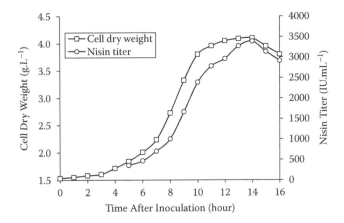

FIGURE 10.1
Biomass and nisin production in batch fermentation carried out in a 1-L benchtop continuously stirred tank reactor (CSTR). (From Wu, Z., Wang, L., Jing, Y., Li, X., and Zhao, Y. 2009. *Applied Biochemistry and Biotechnology* 152, 372–82.)

There are many reasons that batch fermentation is chosen by the industry over continuous fermentation for nisin production in addition to the simplicity of batch operation. For instance, continuous fermentation requires simultaneous cell growth and nisin production. However, the conditions for maximum cell growth, maximum nisin productivity, and maximum nisin bioactivity retention are different. The optimal pH for cell growth is between pH 6 and 7 (Hurst and Dring, 1968; Wu et al., 2009) and preferably between pH 6.25 and 6.5; the optimal pH for nisin production is between pH 5.5 and 6; and the optimal pH for the retention of nisin activity is between pH 2.5 and 3. Continuous operation has little advantage over batch-mode operation unless the culture can sustain a high dilution rate (the ratio of flow of the product stream to the volume of the culture in the reactor). As a general rule, the maximum dilution rate is determined by the maximum specific growth rate of the bacterium. To operate the reactor with a dilution rate higher than the maximum specific growth rate, cell recycling or cell immobilization is required. There is a problem, however: At the optimal growth pH, a large amount of nisin is adsorbed at the surface of the cell. Hurst and Dring (1968) found that at pH 6.8, more than 80% of the nisin produced is bound to the cells. To collect the nisin from the cell surface, one needs to drop the pH of the broth to below pH 3 and heat it to over 90°C (explained further in the next section), after which the cells are killed. Not surprisingly, attempts to run nisin fermentation continuously with immobilized cells or cell recycling resulted in little improvement (Chinachoti et al., 1998; Parente and Ricciardi, 1999) over batch-mode fermentation.

In addition to the problems related to the continuous production and recovery of nisin described above, there is yet another practical difficulty that makes continuous fermentation less desirable: the assay of nisin. Currently, there is no reliable method for real-time analysis of the nisin titer in the fermentation broth that can be applied to large-scale manufacturing. The most-used techniques for nisin quantification are based on the measurement of the inhibition of growth produced in a sensitive microorganism, either in a liquid or solid medium (Parente et al., 1995; Wolf and Gibbons, 1996; Cabo et al., 1999; Wu and Li, 2007). The time required to finish an assay varies from 4 to 6 hr (liquid medium), to 12 to 24 hr (solid medium, such as the agar diffusion method (Hirsch, 1950)). This makes the continuous fermentation for a prolonged period of time a significant challenge at large scale.

10.2 Downstream Process Design

As mentioned above, commercial nisin products contain approximately 2.5% pure nisin by weight (most national standards demand no less than 2.25% pure nisin). This is because pure nisin has very high potency and is not suitable for use in food processing. This specification means that nisin manufacturing does not require sophisticated purification, as long as all of the ingredients used in the medium are of food grade or generally regarded as safe (GRAS). Of course, there are various schemes for producing nisin with higher purity, such as using expanded bed ion exchange (Cheigh et al., 2004) or immunoaffinity (Suarez et al., 1997; Prioult et al., 2000) or other chromatography methods. However, these methods increase the production cost of nisin and can introduce compounds that may be of regulatory concern for food applications and are not environmentally friendly.

Conventionally, nisin is recovered by foam fractionation, followed by precipitation. A 1963 United States patent (Hall, 1963) describes the process as follows:

> "Thus the nisin may be recovered by passing air through the nisin-containing culture medium, if desired after the addition of a small quantity, not exceeding 0.1% by weight, of a surface active agent, collecting the foam, adjusting the pH to 2.5, adding 25% of sodium chloride, recovering the precipitated solid by centrifugation and then freeze drying."

The reason that nisin can be recovered by foam fractionation is that the nisin molecule has amphiphilic character and adsorbs to hydrophobic surfaces such as the surface of an air bubble in liquid. However, for nisin itself to support a stable foam, a concentration of at least 7.5 g.L^{-1} is required (Hirsch, 1950). This is much higher than the actual nisin concentration in fermentation broth, which is typically in the range of 10 to 100 mg.L^{-1} (Saavedra and Sesma, 2011). Thus, the foam fractionation of nisin relies on the coexistence of other surface-active materials. Whether a surface-active agent from an external source is needed depends on the foam stability of the fermentation broth, which in turn depends on the composition of the broth. For example, Khaskova and Kruglyakov (1995) reported that in the absence of an added foam-forming agent, no stable foam can be produced from the fermentation broth, while Liu et al. (2010) were able to extract nisin directly from the fermentation broth by an integrated reactor and foam fractionator.

The simple process described above, although effective, produces low-quality products by modern commercial standards due to the apparent lack of a unit operation for the removal of the bacteria and possibly other solids from the broth. The process is improved by the joint effort of Tianjin Kangyi Bioengineering Co., Ltd. and the Department of Bioengineering, Hebei University of Technology, Tianjin, China, by the introduction of a microfiltration unit operation before foam fractionation (Wu et al., 2008). A flow diagram of the unit operations involved in the modified process is given in Figure 10.2.

- A. *Fermentation, acidification, and sterilization.* Due to the lack of an online nisin titre assay method, fermentation is regulated by monitoring the growth of the bacterium. Mechanical stirring is required

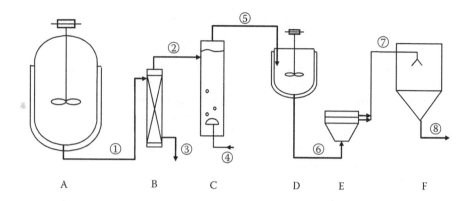

FIGURE 10.2
Process flow diagram of the manufacture of nisin. Unit operations: A: Fermentation, acidification, and sterilization; B: microfiltration; C: foam fractionation; D: precipitation; E: centrifugation; F: spray drying. Material flows: ①: sterilized fermentation broth; ②: permeate; ③: cell debris; ④: compressed clean air; ⑤: foamate; ⑥: precipitate soliquid; ⑦: concentrated precipitate soliquid; ⑧: final product.

to keep the medium homogeneous. Starting from the beginning of the exponential growth phase, the broth may become quite foamy. However, the use of anti-foam at this stage is restricted because too much anti-foam prohibits foam fractionation. Although it is reported that aeration can potentially slightly improve nisin productivity (Desjardins et al., 2001), this is not employed to avoid excessive foaming. At the end of the exponential growth phase, the broth is acidified to pH 2 to 3 by the addition of HCl solution and then is heated to 90°C using steam. There are two purposes for this: (1) to kill the nisin-producing bacteria and release the nisin previously adsorbed to the cell surface, and (2) to deactivate the proteolytic enzymes that could digest nisin. This process lasts about 30 min, after which the broth is cooled to 40°C to 50°C.

- B. *Microfiltration.* Debris of the producer microbe and undigested fermentation media components are then removed from the broth by cross-flow micro-filtration at 0.1 MPa. A ceramic membrane with 0.2 μm pore size is used (in comparison to the cell size of 0.5 to 1.5 μm). A ceramic membrane is used instead of polymeric hollow fiber membranes made of synthetic materials such as polyolefin because the latter normally have hydrophobic surfaces, which completely adsorb nisin (Chinachoti et al., 1998).

As the filtration proceeds, the retentate becomes more and more concentrated with cell debris and other solid materials from the broth. This debris and solid particles adsorb a certain amount of nisin. To maximize the recovery of nisin from this unit operation, at the final stage some pH 2.5 HCl solution with 0.1% Tween™-80 is added to the microfiltration retentate to wash off the adsorbed nisin. This approach can increase the nisin recovery from 68% to 85% (Wu et al., 2008), and even 90% at the industrial scale. Around 10% nisin activity is lost in the retentate, and 3% is lost due to nisin denaturation or adsorption to the membrane and piping (Wu et al., 2008). The filtration permeate has a translucent light brown color while the broth is opaque due to the existence of biomass and other solid particles. The permeate can be stored in an intermediate container but can also be pumped directly to the next step of unit operation, foam fractionation, where the foam columns also serve as storage tanks for the permeate.

- C. *Foam fractionation.* Foam fractionation is operated in semi-batch mode (i.e., batch with respect to the liquid and continuous with respect to the gas), which is compressed air cleaned via deep-bed filtration to remove dust and other microparticles. Multiple foam columns can run in parallel to increase the throughput, and the throughput is designed to match the batch-fermentation cycle but still allow sufficient time for cleaning-in-place (CIP).

    Typical of a semi-batch foam fractionation operation, the nisin concentration decreases with time. Ideally, foam fractionation should stop when the nisin concentration in the residual liquid drops to a certain low level. Unfortunately, as mentioned above, a real-time, online nisin concentration measurement method is unavailable. Hence, termination of the foam fractionation operation is based on the stability of the foam in the column (foam columns are made of transparent acrylic glass for easy observation of the condition of the foam within the columns). When the nisin (and other proteins) concentration is too low, a stable foam is not manifest and foam fractionation must stop. No matter what the initial nisin concentration in the feed solution is, the final residual concentration is approximately 50 IU.mL^{-1}.

    The foam fractionation unit operation can concentrate the filtrate to reduce its volume by a factor of approximately 10 with a recovery rate of 80–85%. There is about 10% loss of nisin activity in the foam fractionation unit operation. Part of this loss is due to adsorption of nisin onto the surfaces of the column and pipes and part of it is due to denaturation in foam (Clarkson et al., 1999a, 1999b).

- D. *Precipitation.* The foamate from the foam fractionation column is directed into the precipitation tank. A selected anti-foam agent (usually a combination of several different food-grade ingredients) is used together with a slow mechanical impeller to collapse the foam as soon as it reaches the precipitation tank to preserve the nisin activity. Prolonged exposure of nisin to the air-liquid interface can cause denaturation and loss of bioactivity of the nisin molecule. When one batch of foam fractionation finishes, food-grade ammonium sulfate (note that ammonium sulfate is GRAS and is also used as an acidity regulator in flours and breads, etc.) is added to the foamate at a concentration of 50% (w/w) to "salt out" nisin. Thanks to foam fractionation, the amount of ammonium sulfate required for nisin precipitation can be reduced by as high as 90%. For example, when 20 tonnes of permeate is concentrated to 2 tonnes by foam fractionation, the ammonium sulfate consumption is reduced from 10 tonnes to 1 tonne only. This also significantly reduces the load of the subsequent centrifugation operation and wastewater treatment.

- E. *Centrifugation* and F. *spray drying.* Further downstream processing is straightforward. The precipitate soliquid (i.e., solid protein particles dispersed in liquid) is transferred to an ultra-high-speed centrifuge where the solid is separated. The solid is resuspended in an NaCl solution and sent to a spray dryer to produce a fine powder. The nisin potency of the powder is measured and NaCl is added accordingly to meet regulatory requirements.

The foam fractionation unit operation can be replaced by ultra-filtration, but the capital cost and operational cost are much higher. On the positive side, ultra-filtration gives a higher nisin purity, but this is unnecessary because a purity of only 2.25% is required. Therefore, foam fractionation provides a more commercially viable alternative to ultra-filtration in the manufacture of nisin.

10.3 Foam Fractionation in Nisin Production

In the above, a brief description was given to show how foam fractionation is employed in the downstream processing of nisin production. In this section, the design of the foam fractionation unit operation is discussed in more detail.

10.3.1 Design of the Liquid Pool Depth

As described in the previous section, foam fractionation is used to concentrate the nisin-containing feed solution. To obtain a high enrichment ratio, a low superficial gas velocity (1 to 2 mm.s^{-1}) and thus a relatively dry foam (ε = 0.01 to 0.02; i.e., 1% to 2%) are used. For such dry foam, once the bubbles leave the liquid and enter the foam layer, the small amount of liquid contained in the Plateau borders is not enough to cause significant further adsorption onto the gas–liquid interface. Thus, it is assumed that the adsorption occurs only while the bubbles rise in the liquid pool, and the extent of adsorption on to the surface of a bubble is determined by the gas residence time in the liquid, which in turn depends on the depth of the liquid pool and bubble rise velocity. This assumption has been adopted by many other researchers who study foam fractionation of proteins (e.g., Bhattacharjee et al., 1997, 2001; Maruyama et al., 2007; Maruyama et al., 2008).

The gas residence time, τ, is related to column dimensions and operating parameters via the following equation:

$$\tau = \frac{L}{v_g} \tag{10.1}$$

where L is the depth of the bubbly liquid (i.e., the distance from the gas sparger to the foam/liquid interface); v_g is the absolute gas velocity in a Eulerian frame of reference (i.e., relative to a stationary observer); and

$$v_g = \frac{j_g}{\varepsilon_g} \tag{10.2}$$

where j_g is the superficial gas velocity and ε_g is the gas volumetric fraction or gas hold-up in the bubbly liquid. Using Equation 10.2, Equation 10.1 can equivalently be expressed as

$$\tau = \frac{\varepsilon_g L}{j_g} = \frac{\varepsilon_g V_l}{Q_g} \tag{10.3}$$

where V_l is total volume of the gas and liquid in the liquid pool of the foam column, and Q_g is the volumetric gas flowrate. Because all the parameters on the right-hand side of Equation 10.3 can be readily measured, the gas residence time can be determined.

Because the superficial gas velocity employed in the foam fractionation of nisin is small, the gas fraction in the bubbly liquid is also low ($\varepsilon_g < 0.06$); the hindering effect discussed in Chapter 5 is insignificant so the absolute gas velocity is more or less the same as the terminal bubble rising velocity, v_∞. If the bubble size can be maintained relatively constant, Equation 10.2 shows a linear relationship between ε_g and j_g, where the slope is the reciprocal of the bubble terminal velocity, that is,

$$\varepsilon_g = \frac{1}{v_\infty} j_g \tag{10.4}$$

Figure 10.3 shows such a relationship given by a nisin-containing microfiltration permeate in a column with 0.05 m internal diameter (Li, 2008). The total protein content in the liquid phase was 2.4 g.L^{-1}. Bubbles were

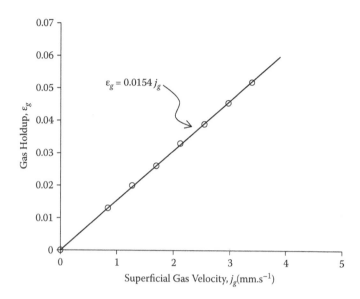

FIGURE 10.3
The relationship between gas hold-up in the liquid pool and superficial gas velocity given by a nisin-containing microfiltration permeate.

generated from a perforated plate sparger with 150-μm pore diameter and 4 mm pore distance. The nisin titer and bubble size in this specific case were not reported, but from the slope of the line it can be seen that $v_\infty = 0.065$ m.s^{-1}.

This above assumption that adsorption occurs only in the bubbly liquid is compatible with another assumption that the nisin concentration in the interstitial liquid (i.e., the liquid in the Plateau borders of the foam) is equal to that in the bubbly liquid beneath the foam, which enables the estimation of surface excess, Γ, by measurement of the foamate concentration using the batch foam fractionation technique (Section 6.3), that is,

$$\Gamma = \frac{(C_P - C_B)V_P r_{32}}{3V_{foam}} \tag{10.5}$$

where C_P is the nisin concentration in the foamate (collapsed foam liquid); C_B is the nisin concentration in the bubbly liquid; V_P is the volume of the foamate (liquid only) and V_{foam} is the total volume of the foam before collapse (i.e., gas volume + liquid volume; however, because the liquid fractionation is very low, it is approximately the same as the gas volume in the foam); and r_{32} is the Sauter mean bubble radius. This method was first used by Lemlich (1968) and adopted by many others (e.g., Bhattacharjee et al., 1997; Maruyama et al., 2000; Li et al., 2011). Because the bubbles in the foam are polyhedral rather than spherical, the bubble radius at the gas–liquid interface can be

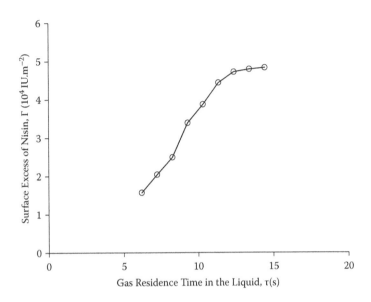

FIGURE 10.4
Adsorption of nisin onto bubble surface as a function of bubble residence time in the liquid.

used for an estimation. Figure 10.4 shows some data of nisin adsorption at different gas residence times. The nisin titer in the bubbly liquid was 1918 $IU.mL^{-1}$. The data were obtained in a benchtop foam column with 25 mm internal diameter and 75 mm height. The gas sparger was made from a piece of filter cloth (Li, 2005). The bubble diameter generated from the system was $r_{32} = 0.28$ mm, measured by photography. The terminal bubble velocity was found to be around 0.0485 $m.s^{-1}$, and it was verified by the dynamic gas disengagement method; i.e., at steady state, the gas supply was suddenly cut off and the time it took for the bubbles to rise to the top of the liquid pool was measured (Sriram and Mann, 1977). The gas residence time was then calculated from Equation 10.1. The surface excess of nisin, measured in IU per square meter of bubble surface area, was calculated from Equation 10.5 by keeping the superficial gas velocity constant and varying only the depth of the bubbly liquid.

Figure 10.4 shows that, in order to guarantee maximum surface excess, a liquid pool gas residence time of 12.5 seconds (s) is required. That being said, nisin is not the only surface-active substance in the fermentation broth, and it is not even the main foam-stabilizing agent. In the filtration permeate provided by Tianjin Kangyi Bioengineering Co., Ltd., the total protein concentration was 2.4 $g.L^{-1}$ while the nisin concentration was typically around 50 to 100 $mg.L^{-1}$ (1 mg pure nisin has a potency of 40,000 IU). For such a complex system, a mechanistic model of nisin adsorption could be prohibitively complicated, although a phenomenological model can be built (Li, 2005;

Wu et al., 2008). Instead of using a model to scale-up the foam fractionation column for industrial use, the gas residence time is used as the sole parameter to determine the depth of the bubble liquid of a foam fractionator. For example, if the bubbles have a terminal velocity of 0.18 m.s⁻¹, the liquid depth then must be around 2.3 m to allow for a gas residence time of 12.5 s. A smaller bubble size is favorable toward interfacial adsorption in the liquid pool, but at the same time it makes the foam wetter and reduces the enrichment. If the liquid pool depth is limited, internals can be used to engender different bubble sizes in the liquid pool and the foam layer to have more adsorption and lower liquid fraction at the same time.

10.3.2 Design of the Foam Height

The foam stabilized by the nisin fermentation broth is relatively stable although some internal bubble coalescence still occurs. The extent of this can be examined by changing the height of the foam layer and measuring the superficial liquid velocity at a constant superficial gas velocity. Such experiments were performed in the same column as used to generate Figure 10.3, and a typical set of data collected this way is shown in Figure 10.5: The liquid pool depth was 0.8 m and the superficial gas velocity was kept constant at 4.24 mm.s⁻¹. The foam height was varied from 0.2 to 0.8 m by adding different numbers of 20 cm column sections to the top of the foam column.

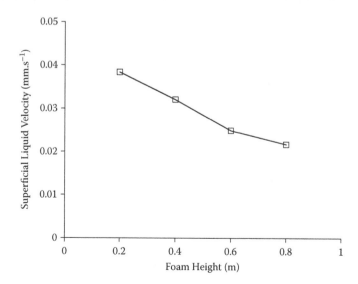

FIGURE 10.5
Superficial liquid velocity given by columns with different foam heights at the same superficial gas velocity. (From Li, X. 2008. Development of a multistage-draining foam fractionator and its application in concentrating nisin from its fermentation broth. Department of Biochemical Engineering, Hebei University of Technology, Tianjin, China.)

At each foam height, the foam was directed via an inverted U-shaped hose to a bucket placed on a digital balance, from where the mass of the foamate was recorded as a function of time. The volumetric liquid flow rate was then calculated from the mass flow rate and the density of the foamate. Some raw data are given in Figure 10.6.

From Figure 10.6 it can be seen that there was a brief unstable period after the gas was switched on due to the collapse of foam at the surface of the foam before the foam completely fills the column. After this induction period, the mass of foamate started to increase linearly with time, indicating a relatively stable foaming period. The superficial liquid velocities presented in Figure 10.5 were all obtained from stable foaming periods. It can be seen that as the total foam height increased, the superficial liquid velocity decreased due to internal coalescence of the foam and consequently an increase in average bubble size. A result of the internal coalescence is "internal reflux" (Lemlich, 1968; Stevenson et al., 2008), which increases the liquid fraction in the lower section of the foam layer (Figure 10.7) and potentially enhances interfacial adsorption.

Figure 10.7 shows the axial liquid fraction profile in a foam column with different total foam heights. The data were obtained by suddenly isolating the different column sections and weighing the foam in each section. This method is simple and has a low spatial resolution (0.2 m); but because it is a direct measurement, the data are reliable. It can be seen from Figure 10.7 that as the foam height was increased from 0.2 to 0.6 m, the liquid fraction

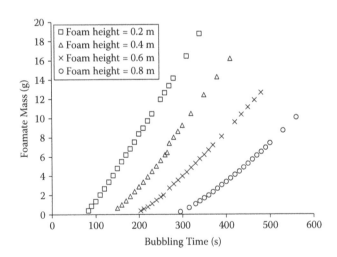

FIGURE 10.6
The raw data from which the superficial liquid velocities in Figure 10.5 were calculated. (From Li, X. 2008. Development of a multistage-draining foam fractionator and its application in concentrating nisin from its fermentation broth. Department of Biochemical Engineering, Hebei University of Technology, Tianjin, China.)

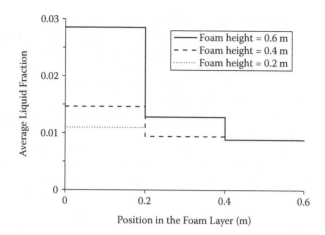

FIGURE 10.7
Effect of total foam height on the axial liquid fraction profile.

at the bottom section of the foam layer increased from 0.011 to 0.029, due to internal reflux. There was no significant difference in the top section of the foam layer when the total foam height was varied, probably due to the fact that the foam was already very dry and the foam drainage rate was very low. A significant enhancement in enrichment was not observed either. However, one cannot say the same for larger columns with further increases in foam height. Because the foam columns are scaled-up on the basis of a superficial gas velocity, an increase in the absolute foam height increases the bubble residence time in the foam column and allows more bubbles to coalesce. In the industrial-sized foam fractionation columns used by Tianjin Kangyi Bioengineering Co., Ltd., the total foam height is 1.7 m, and the liquid fraction at lower sections of the foam layer is enhanced to a much larger extent. At this scale, the foam height does have some, albeit small, impact on the enrichment and recovery of the process. Its effect is dependent on the foam stability, which may vary from batch to batch due to different fermentation conditions. (There are many noncontrollable variables in fermentation. Even beers and wines have different qualities from batch to batch.) Thus, the design of foam layer height relies largely on experience. Generally speaking, a higher foam layer gives higher enrichment at the cost of lower recovery rate.

10.3.3 Design of Other Features

The liquid pool and the foam layer are the two main parts of a foam fractionation column. In addition to the design of the sections, the sparger and the foam collection method design also demand some consideration.

As discussed in Chapter 7, for solutions with a high foaming tendency such as the nisin broth, spargers that generate relatively large bubbles are better suited. For columns with a large internal diameter, uniform distribution of the gas bubbles across the cross-sectional area of the column must also be considered, in order to avoid internal circulation within the liquid pool, which reduces average gas residence time. A lower pressure drop across the sparger is desirable from an energy efficiency point of view, but a certain pressure drop is required to prevent weeping of liquid as described in Section 7.2. The construction material of the sparger must be able to withstand the corrosive nature of the fermentation broth.

Because the liquid fraction used in nisin foam fractionation is very low, the foam does not flow under gravity. A closed conduit is required to collect the foam, and the foam flows in the conduit due to the pressure originating from the buoyancy force of the bubbles at the foam-liquid interface (Li et al., 2012). A closed conduit is also required to meet hygiene requirements.

It is noted that, although nisin has antimicrobial power, it is only effective against G+ bacteria but has no effect on G− bacteria or fungi, etc. Consequently, foam fractionation columns, including the sparger and the foam collection conduit, must be cleaned regularly to prevent contamination.

10.4 Confounding Factors

As discussed above, in order to obtain a high enrichment, the superficial gas velocity used in the foam fractionation of nisin is very low. Consequently, the foam fractionation columns have very low throughput. A column with 0.5 m diameter and 4.0 m height can only process 0.45 m³ liquid feed in every 8-hour shift, including the column filling and discharging, CIP, etc., which takes about an hour. The net throughput is approximately 0.072 m³ feed per cubic meter column volume per hour. A 20-tonne fermenter requires a matrix of sixteen parallel columns operating throughout two 8-hour shifts. It is desirable to intensify the process and improve the throughput without increasing the capital cost or operational cost. However, this is very difficult, and no real breakthrough has ever been made. Multistage foam fractionation (Schügerl, 2000; Boonyasuwat et al., 2003; Darton et al., 2004; Boonyasuwat et al., 2005, 2009) or foam fractionation with external reflux (Lemlich and Lavi, 1961; Lee and Maa, 1986; de Lucena et al., 1996; Yamagiwa et al., 2001) cannot be used because nisin foam has high stability and requires anti-foam to completely collapse it. The resultant foamate would contain anti-foam, which would preclude further foaming; nor can it be returned to the foam column as external reflux or it will collapse the foam there. Although the foam can be collapsed by other methods, repeated foaming could cause severe denaturation of nisin. Promoting drainage is also quite difficult because the current scheme uses a very dry foam. Different foam layer internals were tested at Tianjin Kangyi Bioengineering Co., Ltd. in an attempt to enhance drainage or promote internal reflux, but with limited success. Preliminary experimental results showed that the use of internals that cause successive contraction and expansion (see Section 9.6) can enhance drainage to some extent, but the device occupies the whole cross-sectional area of the column and interferes with CIP, especially when multiple units of the foam riser are installed.

To summarize, foam fractionation is a commercially viable alternative to ultra-filtration in its current state. Further intensification of the process is desirable, but very difficult to employ in the context of nisin production. The most straightforward way to improvement is by using a lower superficial gas velocity and/or a relatively large bubble size.

Nomenclature

Roman

C_{bulk}	Nisin titer in the liquid pool	$[IU.m^{-3}]$
$C_{foamate}$	Nisin titer in the foamate	$[IU.m^{-3}]$
j_g	Superficial gas velocity	$[m.s^{-1}]$
L	Liquid pool depth	$[m]$
Q_g	Volumetric gas flowrate	$[m^3.s^{-1}]$
r_{32}	Sauter mean bubble radius	$[m]$
v_g	Absolute gas velocity	$[m.s^{-1}]$
v_∞	Bubble terminal velocity	$[m.s^{-1}]$
V_{foam}	Volume of foam (gas + liquid)	$[m^3]$
$V_{foamate}$	Volume of foamate	$[m^3]$
V_l	Total volume of liquid and gas in the liquid pool	$[m^3]$

Greek

ε_g	Volumetric gas fraction in the liquid pool	$[-]$
τ	Gas residence time in the liquid pool	$[s]$

References

Abee, T., and Delves-Broughton, J. 2003. Bacteriocins—Nisin, in *Food Preservatives*, Springer, US.

Bertrand, N., Fliss, I., and Lacroix, C. 2001. High nisin-Z production during repeated-cycle batch cultures in supplemented whey permeate using immobilized *Lactococcus lactis* UL719, *International Dairy Journal*, 11, 953–960.

Bhattacharjee, S., Kumar, R., and Gandhi, K.S. 1997. Prediction of separation factor in foam separation of proteins, *Chemical Engineering Science*, 52, 4625–4636.

Bhattacharjee, S., Kumar, R., and Gandhi, K.S. 2001. Modelling of protein mixture separation in a batch foam column, *Chemical Engineering Science*, 56, 5499–5510.

Boonyasuwat, S., Chavadej, S., Malakul, P., and Scamehorn, J.F. 2003. Anionic and cationic surfactant recovery from water using a multistage foam fractionator, *Chemical Engineering Journal*, 93, 241–252.

Boonyasuwat, S., Chavadej, S., Malakul, P., and Scamehorn, J.F. 2005. Surfactant recovery from water using a multistage foam fractionator. Part I. Effects of air flow rate, foam height, feed flow rate and number of stages, *Separation Science and Technology*, 40, 1835–1853.

Boonyasuwat, S., Chavadej, S., Malakul, P., and Scamehorn, J.F. 2009. Surfactant recovery from water using a multistage foam fractionator: Effect of surfactant type, *Separation Science and Technology*, 44, 1544–1561.

Cabo, M.L., Murado, M.A., Gonzalez, M.P., and Pastoriza, L. 1999. A method for bacteriocin quantification, *Journal of Applied Microbiology*, 87, 907–914.

Cheigh, C.I., Kook, M.C., Kim, S.B., Hong, Y.H., and Pyun, Y.R. 2004. Simple one-step purification of nisin Z from unclarified culture broth of *Lactococcus lactis* subsp. *lactis* A164 using expanded bed ion exchange chromatography, *Biotechnology Letters*, 26, 1341–1345.

Chinachoti, N., Endo, N., Sonomoto, K., and Ishizaki, A. 1998. Bioreactor systems for efficient production and separation of nisin Z using *Lactococcus lactis* IO-1, *Journal of the Faculty of Agriculture - Kyushu University*, 43, 421–436.

Clarkson, J.R., Cui, Z.F., and Darton, R.C. 1999a. Protein denaturation in foam. I. Mechanism study, *Journal of Colloid and Interface Science*, 215, 323–332.

Clarkson, J.R., Cui, Z.F., and Darton, R.C. 1999b. Protein denaturation in foam. II. Surface activity and conformational change, *Journal of Colloid and Interface Science*, 215, 333–338.

Darton, R.C., Supino, S., and Sweeting, K.J. 2004. Development of a multistaged foam fractionation column, *Chemical Engineering and Processing*, 43, 477–482.

Davies, E.A., Bevis, H.E., Potter, R., Harris, J., Williams, G.C., and Delves-Broughton, J. 1998. Research note: The effect of pH on the stability of nisin solution during autoclaving, *Letters in Applied Microbiology*, 27, 186–187.

de Lucena, S.L., Miranda, E.A., and Santana, C.C. 1996. The effect of external reflux on the foam fractionation of proteins, *Applied Biochemistry and Biotechnology*, 57-588, 57–65.

Delves-Broughton, J. 2005. Nisin as a food preservative, *Food Australia*, 57, 525–527.

Delves-Broughton, J., Blackburn, P., Evans, R.J., and Hugenholtz, J. 1996. Applications of the bacteriocin, nisin, *Antonie Van Leeuwenhoek*, 69, 193–202.

Desjardins, P., Meghrous, J., and Lacroix, C. 2001. Effect of aeration and dilution rate on nisin Z production during continuous fermentation with free and immobilized *Lactococcus lactis* UL719 in supplemented whey permeate, *International Dairy Journal*, 11, 943–951.

EFSA, 2006a. Opinion of the scientific panel on food additives, flavourings, processing aids and materials in contact with food on a request from the commission related to: The use of nisin (E 234) as a food additive, *The EFSA Journal*, 314, 1–16.

EFSA, 2006b. Opinion of the scientific panel on food additives, flavourings, processing aids and materials in contact with food on a request from the commission related to: The use of nisin (E 234) as a food additive Question number EFSA-Q-2005-031b, *The EFSA Journal*, 314b, 1–8.

Hall, R.H.1963. Production of Nisin, United States Patent US3093551.

Hirsch A. 1950. The assay of the antibiotic nisin, *Microbiology*, 4, 70–483.

Hurst, A., and Dring, G.J. 1968. The relation of the length of lag phase of growth to the synthesis of nisin and other basic proteins by *Streptococcus lactis* grown under different cultural conditions, *Journal of General Microbiology*, 50, 383–390.

Khaskova, T.N., and Kruglyakov, P.M. 1995. Regularities in the concentration and separation of surfactants in a foam, *Russian Chemical Reviews*, 64, 235–247.

Lee, J., and Maa, J.R. 1986. Separation of a surface active solute by foam fractionation, *International Communications in Heat and Mass Transfer*, 13, 465–473.

Lemlich, R. 1968. Adsorptive bubble separation methods: Foam fractionation and allied techniques, *Industrial and Engineering*, 60, 16–29.

Lemlich, R., and Lavi, E. 1961. Foam fractionation with reflux, *Science*, 134, 191.

Li, X., 2005. Adsorption of nisin onto bubble surfaces in a batch foam fractionator: Kinetics study, Department of Bioengineering, Hebei University of Technology, Tianjin, China.

Li, X., 2008. Development of a multistage-draining foam fractionator and its application in concentrating nisin from its fermentation broth, in Department of Biochemical Engineering, Hebei University of Technology, Tianjin, China.

Li, X., Evans, G.M., and Stevenson, P. 2011. Process intensification of foam fractionation by successive contraction and expansion, *Chemical Engineering Research & Design*, 89, 2298–2308.

Li, X., Evans, G.M., and Stevenson, P. 2012. Remarks on approaches that relate foam drainage rate to bubble terminal velocity, *International Journal of Multiphase Flow*, 42, 24–28.

Liu, W., and Hansen, J.N. 1990. Some chemical and physical properties of nisin, a small-protein antibiotic produced by *Lactococcus lactis*, *Applied and Environment Microbiology*, 56, 2551–2558.

Liu, W., Zheng, H., Wu, Z., and Wang, Y. 2010. Effects of pH profiles on nisin fermentation coupling with foam separation, *Applied Microbiology and Biotechnology*, 85, 1401–1407.

Lück, E., and Jager, M. 1997. Nisin, in *Antimicrobial Food Additives*, Springer-Berlin/Heidelberg.

Lv, W., Cong, W., and Cai, Z. 2004a. Nisin production by *Lactococcus lactis* subsp. *lactis* under nutritional limitation in fed-batch culture, *Biotechnology Letters*, 26, 235–238.

Lv, W., Cong, W., and Cai, Z. 2004b. Improvement of nisin production in pH feedback controlled, fed-batch culture by *Lactococcus lactis* subsp. *lactis*, *Biotechnology Letters*, 26, 1713–1716.

Lv, W., Zhang, X., and Cong, W. 2005. Modelling the production of nisin by *Lactococcus lactis* in fed-batch culture, *Applied Microbiology and Biotechnology*, 68, 322–326.

Maruyama, H., Seki, H., Matsukawa, Y., Suzuki, A., and Inoue, N. 2008. Adsorption behavior of bisphenol-A and diethyl phthalate onto bubble surface in nonfoaming adsorptive bubble separation, *Chemical Engineering Journal*, 141, 112–118.

Maruyama, H., Seki, H., Suzuki, A., and Norio, I. 2007. Batch foam separation of a soluble protein, *Water Research*, 41, 710–718.

Maruyama, H., Suzuki, A., and Seki, H. 2000. Adsorption of water-soluble proteins onto bubbles in continuous foam separation, *Journal of Colloid and Interface Science*, 224, 76–83.

Parente, E., Brienza, C., Moles, M., and Ricciardi, A. 1995. A comparison of methods for the measurement of bacteriocin activity, *Journal of Microbiological Methods*, 22, 95–108.

Parente, E., and Ricciardi, A. 1999. Production, recovery and purification of bacteriocins from lactic acid bacteria, *Applied Microbiology and Biotechnology*, 52, 628–638.

Penna, T.C., and Moraes, D.A. 2002. Optimization of nisin production by *Lactococcus lactis*, *Applied Biochemistry and Biotechnology*, 98-100, 775–789.

Pongtharangku, T., and Demirci, A. 2007. Online recovery of nisin during fermentation and its effect on nisin production in biofilm reactor, *Applied Microbiology and Biotechnology*, 74, 555–562.

Prioult, G., Turcotte, C., Labarre, L., Lacroix, C., and Fliss, I. 2000. Rapid purification of nisin Z using specific monoclonal antibody-coated magnetic beads, *International Dairy Journal*, 10, 627–633.

Saavedra, L., and Sesma, F., 2011. Purification techniques of bacteriocins from lactic acid bacteria and other Gram-positive bacteria, in *Prokaryotic Antimicrobial Peptides*, Springer, New York.

Schügerl, K. 2000. Recovery of proteins and microorganisms from cultivation media by foam flotation, in *Advances in Biochemical Engineering/Biotechnology*, Springer-Verlag, Berlin/Heidelberg.

Sriram, K., and Mann, R. 1977. Dynamic gas disengagement: A new technique for assessing the behaviour of bubble columns, *Chemical Engineering Science*, 32, 571–580.

Stevenson, P., Li, X., and Evans, G.M. 2008. A mechanism for internal reflux in foam fractionation, *Biochemical Engineering Journal*, 39, 590–593.

Suarez, A.M., Azcona, J.I., Rodriguez, J.M., Sanz, B., and Hernandez, P.E. 1997. One-step purification of nisin A by immunoaffinity chromatography, *Applied Environmental Microbiology*, 63, 4990–4092.

Van't Hul, J.S., and Gibbons, W.R. 1996. Concentration and recovery of the bacteriocin nisin from *Lactococcus lactis* subsp. *lactis*, *Biotechnology and Applied Biochemistry*, 24, 251–256.

Wolf, C.E., and Gibbons W.R. 1996. Improved method for quantification of the bacteriocin nisin, *Journal of Applied Bacteriology*, 80, 453–457.

Wu, Z., Ji, Y., Guo, Y., and Hu, J. 2008. Application of ceramic membrane filtration to remove the solid in nisin fermentation broth, *International Journal of Food Engineering*, 4, art. 2.

Wu, Z., and Li, X. 2007. Modification of the data-processing method for the turbidimetric bioassay of nisin, *Applied Microbiology and Biotechnology*, 74, 511–516.

Wu, Z., Wang, L., Jing, Y., Li, X., and Zhao, Y. 2009. Variable volume fed-batch fermentation for nisin production by *Lactococcus lactis* subsp. *lactis* W28, *Applied Biochemistry and Biotechnology*, 152, 372–82.

Wu, Z., Zheng, H., and Li, X. 2008. Adsorption of nisin onto bubble surfaces in a batch foam fractionator: Kinetics study, *Journal of Biotechnology*, 136, Supplement, S498.
Yamagiwa, K., Iwabuchi, M., Tsubomizu, H., Yoshida, M., and Ohkawa, A. 2001. Effect of external reflux on foam fractionation of poly(vinyl alcohol), *Journal of Chemical Engineering of Japan*, 34, 126–131.

Index

For Product Safety Concerns and Information please contact our EU
representative GPSR@taylorandfrancis.com Taylor & Francis Verlag GmbH,
Kaufingerstraße 24, 80331 München, Germany

Printed and bound by CPI Group (UK) Ltd, Croydon, CR0 4YY
01/05/2025
01858573-0001